Ioana-Andreia Afilipoie (Rădășanu)
Ioan Gontariu
Alice Roșu

Research on the characteristics of cereals beverage - braga

Ioana-Andreia Afilipoie (Rădășanu)
Ioan Gontariu
Alice Roșu

Research on the characteristics of cereals beverage - braga

LAP LAMBERT Academic Publishing

Impressum / Imprint

Bibliografische Information der Deutschen Nationalbibliothek: Die Deutsche Nationalbibliothek verzeichnet diese Publikation in der Deutschen Nationalbibliografie; detaillierte bibliografische Daten sind im Internet über http://dnb.d-nb.de abrufbar.
Alle in diesem Buch genannten Marken und Produktnamen unterliegen warenzeichen-, marken- oder patentrechtlichem Schutz bzw. sind Warenzeichen oder eingetragene Warenzeichen der jeweiligen Inhaber. Die Wiedergabe von Marken, Produktnamen, Gebrauchsnamen, Handelsnamen, Warenbezeichnungen u.s.w. in diesem Werk berechtigt auch ohne besondere Kennzeichnung nicht zu der Annahme, dass solche Namen im Sinne der Warenzeichen- und Markenschutzgesetzgebung als frei zu betrachten wären und daher von jedermann benutzt werden dürften.

Bibliographic information published by the Deutsche Nationalbibliothek: The Deutsche Nationalbibliothek lists this publication in the Deutsche Nationalbibliografie; detailed bibliographic data are available in the Internet at http://dnb.d-nb.de.
Any brand names and product names mentioned in this book are subject to trademark, brand or patent protection and are trademarks or registered trademarks of their respective holders. The use of brand names, product names, common names, trade names, product descriptions etc. even without a particular marking in this work is in no way to be construed to mean that such names may be regarded as unrestricted in respect of trademark and brand protection legislation and could thus be used by anyone.

Coverbild / Cover image: www.ingimage.com

Verlag / Publisher:
LAP LAMBERT Academic Publishing
ist ein Imprint der / is a trademark of
OmniScriptum GmbH & Co. KG
Heinrich-Böcking-Str. 6-8, 66121 Saarbrücken, Deutschland / Germany
Email: info@lap-publishing.com

Herstellung: siehe letzte Seite /
Printed at: see last page
ISBN: 978-3-659-80969-9

Zugl. / Approved by: Suceava, „Ștefan cel Mare" University of Suceava, Diss., 2015

TABLE OF CONTENT

ABSTRACT

This paper proposed an approach, both from a theoretical point of view as well as a practical one, of the sanogenous potential of a fermented beverages from cereals named braga, which, in Romania, has lost its identity because of the phenomenon of food globalization. Another aim of this work is to describe this liquor from the prism of some parameters such as quality and safety, as well as of those which are responsible for the aromatic profile using instrumental analysis techniques.

Braga is a traditional Turkish fermented beverage mostly appreciated in our country in the past. It is made from a mixture of cereals and is obtained as a result of duble fermentation: the process is initiated by a short alcoholic fermentation, followed by a lactic fermentation during which the majority of flavour compounds is formed.

The first part of the paper addresses the importance and role of fermentative industry and shows the components of braga describing its origins, the variety of recipes and its nutritional virtues, specifying some features of braga products in the region Galați. The second part of the work focuses on appropriate investigations relating to the identification and dosage of specific flavors drink, as well as it analyzes and identifies the trace elements and heavy metals in braga.

The aroma compounds were determined by chromatographic analysis, using a gas chromatograph mass spectrometer equipped with a Shimadzu GC-MS. A number of sixty four flavor compounds have been identified in this beverage most of them being esters. These aroma compounds derived from cereal raw materials, but the majority of them resulted from alcoholic and acetic fermentation.

This beverage was analyzed in term of minerals and heavy metal content using an inductively coupled plasma mass spectrometer equipped with AGILENT 7500 ICP-MS. After having made the analyses high values were recorded for Mg, Na, Al and Ni.

INTRODUCTION

"De gustibus non est disputandum."
"In matters of taste, there can be no disputes."

The role of food for human nutrition has been acknowledged since ancient times, having major implications on the state of health and the labor power, starting with the stage of embryonic development. The beneficial effects of food are being felt as well by and the normal functioning of all the organs and the brain, influencing thus the wellbeing and psychic mood for optimum metabolic processes.

Therefore, food should have nutritional value, providing an optimal ratio in terms of quality and quantity of the basic nutrients - carbohydrates, proteins, lipids and biologically active substances such as vitamins and minerals. Food must be also safe for human consumption, without jeopardizing the human body, microorganisms-free and harmful chemical substances-free, such as heavy metals, mycotoxins, pesticide residues, thus ensuring innocuousness. The absence of the latter one renders a useful product dangerous for consumers, becoming instead a source of diseases, in some cases with lethal effects.

Both in the developed countries and in the developing ones, the phenomenon of food chain globalization has become more and more obvious, through the emergence of new challenges and risks to human health and consumers' interests. Thus, common consumers confused by the variety of colors and flavors on supermarket shelves are not capable any longer of appreciating what is beneficial for their body in terms of nutrition, for them being valid Latin dictum *"De gustibus non est disputandum"*, explained by *"In matters of taste, there can be no disputes"*.

It results from the multilateral analysis of this issue that food industry has increasingly made use of chemicals, thus foods and beverages becoming unnatural, though the big producers are trying to emphasize as much as possible this characteristic of naturalness, by making appeal to different syntheses, complying with the consumers' demands, but not with those of human body cells. Consequently there occurs disequilibrium between the food intake and the dietary needs, the main factor

4

being the deficiency of some nutrients with plastic and energy role, vitamins and mineral salts. All these facts lead to a decrease in the intensity of metabolic processes resulting in the occurrence of human body diseases.

The solution of these problems that seem to put a more and more negative imprint on life is the return to nature and tradition. A simple look into the past reflects the fact that our ancestors left us a food treasure that nowadays is not capitalized properly, but buried deeper by all sort of promotion strategies claiming for being modern and beautiful, but not always beneficial and healthy. Thus, products that once consumers were familiar with and appreciated by for their nutritional value have got lost in the avalanche of fast-food products and carbonated soft drinks whose consumption has registered an ascendant trend in recent years.

Braga, a natural fermented beverage, of oriental origin, obtained from a cereal mixture, was considered in the past a pharma-food due to its nutritional virtues. Its reintroduction in the food chain is a healthy alternative to what the present day market offers, representing a new challenge for potentially producers interested to promoting new products. Braga is a point of interest not only for its nutritional valences, but especially for its refreshing and energizing character, being the healthiest choice for body's nutritional needs, both for children and adults.

Having in view all these considerations, this paper aims at approaching, both from a theoretical point of view and practical one the sanogenous potential of cereal fermented beverage named *braga*, which in Romania has lost its identity because of the embracement of western food proposals, characterized by a wide variety, but not always healthy. At the same time, another aim of this paper is to describe this liquor from the prism of some parameters such as quality and safety, as well as of those which are responsible for the aromatic profile using instrumental analysis techniques. The major aims are:
- to emphasize the importance of fermentative beverages and their benefits on the human body by their consumption;
- to characterize the nutritional potential of cereals used as raw material in the braga production process;

- to characterize the natural beverage - braga, in terms of nutritional composition and sanogenous effects;
- to identify the volatile compounds responsible for product flavor by using gas-chromatography analysis coupled with mass spectrometry detection (GC-MS);
- to determinate the quality and safety parameters of this beverages by applying inductively coupled plasma mass spectrometry (ICP-MS) in terms of heavy metals determination;
- to determinate trace elements as part of the composition of braga using inductively coupled plasma mass spectrometry (ICP-MS).

The paper consists of two parts: the first one includes an overview study on the literature in the field, while the second one presents some personal research, results, discussion and conclusions.

The first chapter, entitled "The Importance of cereal fermented beverage", outlines in a theoretical note the importance and role of fermentation industry, describing the nutritional value of cereals and cereal beverages.

The second chapter, entitled "Braga - source of nutrients in a mixture of fermented cereals" presents the defining elements of this beverage, specifying its origins, displaying the variety of recipes and its nutritional virtues. A series of characteristics of braga made in our country, in the region of Galați, is also described.

The third Chapter and the fourth one are addressing topics based on the research of aromatic profile and content of mineral and heavy metals in braga, presenting the results and discussion of the analyses performed by using instrumental analysis techniques.

CHAPTER 1

THE IMPORTANCE OF CEREAL FERMENTED BEVERAGE

1.1. The role of fermentative industry in obtaining cereal beverages

For centuries, human civilization has appealed to various approaches to prepare and preserve food products, based on empirical knowledge acquired through the experience without theoretical information or scientific basis (Todorov S.D., Holzapfel W.H., 2015).

In this respect, the fermentation industry plays an important role within the food industry, from the fermentative process results a wide range of products, with major implications on human body's health (Dabija Adriana, 2010).

According to some researchers' opinion fermentation remains a safe and convenient process in order to ensure the preservation of products, being used successfully in alternative medicine as well.

The evidence that these products have been used since ancient times in many parts of Asia, Africa, Europe, Middle East and South America, is the food thesaurus inherited from our ancestors. The Egyptians introduced the alcoholic beverages, the nomadic peoples from Central Asia prepared various products such as yogurt, kefir or kumis, the Greeks and Romans prepared products by fermentation of olives, the Germanic tribes were specialized in the fermentation of meat, while the Eskimos focused attention on the fermentation of fish, the ancient Persians were specialists in the preparation of boza (in Romania - braga), a product obtained from cereal fermentation, and the Americans were interested in the products obtained from the fermentation of corn (maize) (Alan J. Marsh et al., 2014; Todorov S.D., Holzapfel W.H., 2015).

Although at that time people had no knowledge of microbiology, this concern being dealt with later in the nineteenth century, these peoples based themselves on their personal experience and managed to preserve a large range of food, most of them being considered pharma-foods.

The fermentation process of food is a very old technology and date from 6000 BC, and its theoretical basis was transmitted from generation to generation within local communities (Caplice E., Fitzgerald G.F., 1999).

If such products were initially processed in small quantities, later on it was necessary to produce them at industrial level, this being the starting point, with satisfactory results, for the commercial segment (Alan J. Marsh et al., 2014).

In addition, developments in science, in general, and in microbiology particulary have grown exponentially by the middle of the nineteenth century. Louis Pasteur put the scientific basis of the fermentation process, so the twentieth century debuted with information about „probiotics", which are live microorganisms that protects the humman body against various diseases such as the gastro-intestinal infections and inflammatory diseases (S.D. Todorov, W.H. Holzapfel, 2015).

Having in view all these aspects, special attention was given to the important role of fermented products both for the life cycle of a product, since it is a safe and economic method of conservation, but especially for consumer's body, defining the sensory characteristics of food products, ensuring high digestibility and having special nutritional values (Alan J. Marsh et al., 2014; Gassem Mustafa A.A., 2002).

Applicability of the fermentation phenomenon is well acknowledged in the manufacture process of beer, alcohol, yeast, wine, milk and acidic dairies, pickles and many other fermented products, based mainly on the activity of the yeast which converts saccharides from raw vegetable materials, even animal ones, into other substances which form the finished product (Dabija Adriana, 2010; Gotcheva Velitchka et al., 2001).

Fermented foods are foods which substrates invaded by beneficial microorganisms that are activated by enzymes, in particular amylases, proteases and lipases that hydrolyze polysaccharides, proteins and lipids, leading to the formation of non-toxic products responsible for the organoleptic characteristics of the product: taste, flavor, texture (Gassem Mustafa A.A., 2002).

Among fermented beverages which present a particular importance are those obtained from cereals, very popular in tropical regions, on the continent of Africa. In most cases the process is initiated by a natural microorganism (such as lactic acid bacteria), which helps the fermentation of grains of wheat, maize, rice, millet, barley, oats, rye, sorghum (Alan J. Marsh et al., 2014).

In the case of cereals, yeasts and lactic acid bacteria are the main agents responsible for the production of fermentation, contributing to imprinting sensory characteristics and safety to the finished product. While yeasts require a carbohydrate substrate to produce mainly ethanol, bacteria need a protein substrate, resulting mainly lactic acid (Todorov S.D., Holzapfel W.H., 2015).

Also, the fermentation produced by yeasts and *lactobacilli* influences the finished product in terms of vitamins, minerals and fats (Gotcheva Velitchka et al., 2001).

Although the beneficial role of yeast from fermented beverages has not completely elucidated, many studies reported that due to the proteolytic and lipolytic activities of yeast, aromatic compounds are obtained, helping bacteria growth by producing amino acids, vitamins and other metabolites, and thus contributing to the final composition of the product, by producing ethanol and carbon dioxide (Alan J. Marsh et al., 2014).

If in the past these beverages with functional characteristics were highly appreciated due to their health benefits, without making studies on their nutritional composition, at present more and more biotechnological methods are being developed in order to promote them.

The positive action of these fermented beverages is described by their content in beneficial microorganisms. Although some of the functional characteristics of these traditional beverages provide the basic nutrition and raw unfermented ingredients, there is evidence that some of them provide also benefits directly by the action of probiotics and microbiota and indirectly by the production of metabolites and breakdown of proteins complexes (Alan J. Marsh et al., 2014).

The beverages produced from fermented cereal have a high mineral content, and are characterized by a low fat content, as compared to diaries; however, they provide only some of the essential amino acids, because some types of cereals lack

9

these substances. These beverage are rich in fiber, minerals, vitamins, flavonoids and phenolic compounds, which act efficient on oxidative stress, hyperglycemia and carcinogenesis (Wang, Chung-Yi, Sz-Jie Wu, 2013).

Fermented cereal beverages have an important role in diseases of the gastrointestinal tract due to the presence of probiotics, helping to reduce diarrhoea and malnutrition caused by unhealthy beverages used by children after lactation. Also, due to their refreshing character, they help improve the wellbeing of the human body (Alan J. Marsh et al., 2014).

1.2. The nutritional valences of cereals - raw material for braga

Cereal grains are basic food for almost the entire population of the world and can be consumed in various forms: raw, ground, under the form of flour or boiled, having an important role in ensuring food safety of the population, due to a complex composition in nutrients. Also they are raw materials for a number of industries such as: drinks, alcohol, starch, beer, dextrin, glucose, by-products being used as raw material in pulp and paper industries, as well as feed for animals.

Some research studies have shown that cereals can be used as a source of energy, for example biogas or bio ethanol production obtained from fermentation (Koehler P., Wieser H., 2013).

Cereal grains are characterized by a high content of nutrients, being the raw materials necesary to manufacture fermented food. Cereals are used to prepare fermented products consumed successfully in almost all part of the world. Grains are cultivated on agricultural areas longer than 73% and constitute the raw materials for the production of over 60% of food and offer a wide range of nutrients such as protein, fiber, vitamins, minerals and energy. This is the premise that most researchers started their study from regarding the nutritional composition of the products obtained by the fermentation of grains (Todorov S.D., Holzapfel W.H., 2015).

According to F.A.O. statistics (Food and Agriculture Organization) in 2012, the cereal production evaluated in 2010 is shown in table 1.1.

The cereal production in 2010

(F.A.O., 2012)

Species	Area cultivated (million ha)	The cereal production (million tonnes)
Corn / Maize	162	844
Rise	154	672
Wheat	217	651
Barley	48	123
Sorgum + Millet	76	85
Oats	9	20
Triticale	4	13
Rye	5	12

Despite their small volume, cereals (wheat, rye, barley, oats, millet, rice, maize, sorghum) have some features very valuable and highly appreciated by humans, this aspect falls them in the group of plants with particular importance for the activity and human existence, since ancient times. Thus, cereals are characterized by:

- a ratio of 1: 6 between proteins and carbohydrates, very favorable to the human body (as compared to the ratio of 1: 3 as in the case of legumes for grain and 1:12 – 1:16 in the case of potato);
- a low moisture content (usually 11-14%), which provides a good preservation for long periods of time and an easy transport over long distances;
- proper preservation of nutrition values, without being change in time, which places them within the building reserve fund for feeding humans and animals in special cases, crises (drought, natural disasters);
- a diversity of their use, being consumed both as food for people, feed for animals and also as raw material for different industries;
- a great ecological plasticity, grain being grown almost all over the world;
- short vegetation period, after harvesting their crops being cultivated successively;

- the presence of a fully mechanized and relatively simple cultivation technology;
- a fasciculated root system, the roots being spread in the soil surface layer from which they extract their nutrients and can enter in rotation with plants that have a deeper root system, which procure nutrients from deeper layers of soil (Ion V., 2010).

The chemical composition of cereal grains varies according to species, variety and plant, the chemical substances being distributed unevenly in terms of quantity and quality in different anatomical structures. The chemical composition of the main cereals grown in Romania is shown in table 1.2.:

Tabel 1.2.

The chemical composition of the main cereals grown in Romania

(Koehler P., Wieser H., 2013)

	Wheat	Rye	Barley	Oat	Corn	Rice	Millet
	g/100g						
Humidity	12,6	13,6	11,3	12,1	13,1	13,0	12,0
Protein	11,3	9,4	8,8	11,1	10,8	7,7	10,5
Fats	1,8	1,7	3,8	2,1	7,2	2,2	3,9
Carbohydrate	59,4	60,3	65,0	62,7	56,2	73,7	68,2
Fiber	13,2	13,1	9,8	9,7	9,8	2,2	3,8
Minerals	1,7	1,9	1,3	2,3	2,9	1,2	1,6
	mg/kg						
Vitamin B_1	4,6	3,7	3,6	4,3	6,7	4,1	4,3
Vitamin B_2	0,9	1,7	2,0	1,8	1,7	0,9	1,1
Nicotinamide (B3)	51,0	18,0	15,0	48,0	24,0	52,0	18,0
Pantothenic acid	12,0	15,0	6,5	6,8	7,1	17,0	14,0
Vitamin B_6	2,7	2,3	4,0	5,6	9,6	2,8	5,2
Folic acid	0,9	1,4	0,3	0,7	0,3	0,2	0,4
Tocopherols	41,0	40,0	66,0	22,0	18,0	19,0	40,0

Cereal grains are a source of *proteins*, especially simple protein. They are distributed over the whole grain, but the concentration differs as follows: germ and the aleuron layer contain more than 30% protein, endosperm approximately 13% and bran 7%. The embryo grains contain considerable quantities of complex proteins -

nucleoproteins and lipoproteins. Prolamines (alcohol-soluble) and glutelins (soluble in alkaline weak) predominate. Proteins (soluble in water) and globulins (soluble in solutions of salts) predominate in the grains of rye. The average protein content of the cereal grains is 8 - 11%, and it varies quantitatively depending on the variety, for example, in the case of wheat grain protein content varies between 6 - 20% (Bârcă Adriana, 2011; Koehler P., Wieser H., 2013).

For most of the cereal grains, the ratio of essential amino acids in protein is different from the optimum one. Depending on the content of essential amino acids, the most precious proteins are those in maize and millet. The wheat grain has a limited content of methionine and lysine, whereas the protein content of the corn grains are limited in tryptophan and lysine. Amino acid composition of proteins, especially of essential amino acid, is a way of appreciating the nutritional quality of traditional grain products, as well as of those obtained from new sources of raw materials (Siminiuc Rodica et al., 2004).

From a quantitative perspective, *carbohydrates* are the most representative among the organic substances from grain. Cereal grains contain 66 - 76% carbohydrates being represented by starch, dextrin, sugars, fibers and gums substances (Koehler P., Wieser H., 2013).

Starch (55-70%) is present under the form of granules in endosperm of cereals and represent the most important chemical component, qualitatively and quantitatively. The starch granules can be distinguished depending on the form, on content of amylose and amylopectin, temperature and gel rate, and during storage, the starch is used as a nutritional substance. Starch grains from millet, rye and wheat have a higher hydration capacity than that of barley and corn grains, thus boiled groats from those grains is characterized by a fine texture. *Carbohydrates* (3%) are represented mainly of sucrose and less by reducing sugars, have a major influence on the ability storage of grain which is reduces when the concentration is higher. *Fibers* are mainly represented by cellulose, hemicellulose and lignin, bringing a significant contribution in terms of nutritional value, their quantity ranging in proportion with the coating of the product. Cellulose and hemicelluloses are found predominantly in skin caryopses. If the unassimilated carbohydrate content increases, especially cellulose, the commercial quality of grain and derived products decreases. *Gums substances* are soluble, their role residing in the ability of imprinting a viscous character to the solutions containing them. The content of these substances in rye and oats grains is about 1.5-2.5% of the grain. Due to these substances, fermented cereal beverages obtained by a simpler technology present a viscous aspect (Bârcă Adriana, 2011; Koehler P., Wieser H., 2013).

Fats are in a ratio of 95% under the form of triglycerides, in seeds and coating, whereas in endosperm, in smaller quantities, under the form of phospholipids and glycolipids, most of them being removed by grinding. The oil from the grain seeds is rich in tocopherols. Free fatty acids (linoleic, linolenic), resulted from the hydrolysis of fats during the storage of cereal products are easily oxidized by the action of oxygen and lipoxygenase to form the hydroperoxide. They are decomposed, causing rancidity of flour and groats. Fat of millet, oats, corn and wheat is relatively easily oxidized. The rye, buckwheat, barley fats are more resistant due to their high content in natural antioxidants - lecithin and vitamins B (Bârcă Adriana, 2011; Dabija Adriana, 2010).

Vitamins play an important role in composition and nutritional value of cereal products and ranges from 1-50% depending on the species and variety, being located mainly in the embryo and aleurone layer. Therefore, in hulled and polished groats and white flour there is a low quantity of vitamins. Cereals are an important source of B vitamins and tocopherols in a concentration of 20 mg/kg are among fat-soluble vitamins which distinguish themselves. Contents of vitamins in the kernels of various cereals is different: 2 - 13 mg/kg vitamin B_1, 0.6 - 2.9 mg/kg vitamin B_2, 4 - 98 mg/kg vitamin PP, 6-10 mg/kg of tocopherol. Grains also contain folic acid, carotene, pyridoxine, biotin, pantothenic acid. A relatively high percentage of vitamins are reported in buckwheat, wheat, rye, barley. *Mineral salts* are important for physiology grain at germination, for yeast nutrition during fermentation and to ensure the optimal pH of enzyme. Mineral content of grain varies between 1 - 2.5%. As compared with other foods, this is an intermediate value with that of milk, meat and vegetables, whose average is 3.5%. Over 90% of all minerals in the grain are in the external layers of grain, representing phosphorus, potassium, calcium, magnesium. Phosphorus is found mostly in the form of phytic acid and its salts – phytates. Under the action of the enzyme phytase, phytates hydrolyzed, releasing phytic acid. Since the phytic acid forms insoluble salts with certain minerals, it cuts down the digestive use of mineral elements. Unequal distribution of mineral salts in grains and their high content in the coating and aleurone layer allows the use of ash content to assess quality of cereal products (Bârcă Adriana, 2011; Dabija Adriana, 2010; Koehler P., Wieser H., 2013).

Enzymes play an important role during the storage of cereals. Among hydrolases, more important is α-amylase, which attacks the bonds inside the macromolecule of starch and β-amylase, which destroys the bonds side, resulting dextrin, maltose and glucose. Other important enzymes are proteases, lipases, oxidases, phosphatases (Bârcă Adriana, 2011).

Cereals are rich in nutrients and provide numerous health benefits, in addition boiled bran provides an excellent substrate for production of probiotics phosphatases (Salovaara H., Simonson L., 2003).

Cereal grains are an important source of proteins, carbohydrates, vitamins, minerals and fibers, with beneficial physiological effect on the organism by stimulating the growth of bacteria of the genus *Lactobacillus* and bifidobacteria present in the colon. The high content of soluble fibers, such as β-glucan and arabinoxylan oligosaccharides such as galacto- and fructo-oligosaccharides and starch grains were found to be the source of the probiotic. *Lactobacillus* strains were recognized as complex microorganisms which need, amino acids, B vitamins, nucleic acids and minerals in order to increase fermentable carbohydrates, and the fermentation of grain is an effective way of obtaining these microorganisms with probiotic properties (Vasudha S., Mishra H. N., 2013).

Barley and oats contain β-glucan, a prebiotic that can help to reduce 20-30% of LDL cholesterol, which leads to a reduction in cardiovascular diseases. Furthermore, β-glucan stimulates the multiplication of bacteria in the colon of humans and animals.

Low glycemic index of barley and oats is beneficial for diabetics, contributing to amend the emulsification level of fats and reducing the lipase activity from gastrointestinal tract (Angelov A. et al., 2006; Jaskari J. et al., 1998).

CHAPTER 2

BRAGA – A SOURCE OF NUTRIENTS IN A MIXTURE OF FERMENTED CEREALS

"The one who has never drunk braga, does not know what he loses, the one who has drunk braga once, he will never forget it!"

Highly appreciated in the past in large "bragagerii" (shops where braga is sold) of the royal avenues until the last slum, braga is an alternative to carbonated drinks, which are most valued by consumers, but have negative effect on the human body.

This beverage, also known as "boza", very popular in Romania during 1880-1940, was brought in the Romanian territory by the Turks, along with many other oriental delights like halva, cataif, but nowadays it is not enjoying the same notoriety, many consumers being reluctant when hearing this term (www.gds.ro, www.121.ro).

2.1. History of the fermented beverage Braga

Braga, known its origins as boza is a highly appreciated beverage due to its nutrient properties. The story of this delicious liquid has its origin in Muslims' history (www.stiri-evenimente.ro). Later it became very popular in the surroundings: Kazakhstan, Kyrgyzstan, Albania, Kosovo, Bulgaria, Macedonia, Montenegro, Bosnia and Herzegovina, Romania, Serbia and Ukraine (www.wikipedia.en). The name of this ancient refreshing beverage comes from the Persian word "buuze darı" which means "millet", used as the main ingredient in some recipes for cooking (www.dambovita.net). In Eastern European countries this traditional Turkish beverage is recognized as "braga", in the Balkans "busa" and in Egypt "bouza" (www.trandfonline.com).

Braga is a refreshing beverage with sweet-sour taste with a deep identity kept for a long time in Mesopotamia and Thrace area, about 8000-9000 years ago. The Greek historian Xenophon records that braga was made in Eastern Anatolia in 401

B.C. and it was kept in clay jars (karase) buried in the ground (LeBlanca J. G., Torodov S.D., 2011).

Braga was a very popular beverage in the Middle Ages in the Balkans, Anatolia and Africa, whose appreciation was conditioned by the religious status of the Sultan of that time (www.hurriyetdailynews.com).

This beverage was taken and spread by the Turks in Central Asia beginning with the tenth century. Later on, it was spread to the Caucasus and the Balkans, and the golden age for this beverage was that of the Ottoman Empire period, when the manufacture of braga become one of the main occupations (LeBlanca J. G., Torodov S.D., 2011).

Until the sixteenth century, it could be drunk for free everywhere, but the emergence of a range of braga so-called "tartar boza", which contained opium, drew the wrath of authorities and it was prohibited by the Sultan Selim II (1566-1574). Then, in the seventeenth century, Sultan Mehmed IV (1648-1687) prohibited the consumption of alcoholic drinks, so they closed all the shops which sold braga because this beverage was included in the alcoholic drinks category (www.stiri-evenimente.ro).

It is believed that the Sultan Fatih Mehmet liked this beverage due to its refreshing properties, and the first store to sell braga was opened in the period of the Sultan Suleyman Magnificent (www.hurriyetdailynews.com).

The Turkish traveler Evliya Çelebi mentioned that in the seventeenth century braga consumption reached its apogee, at that time in Istanbul existing over 300 shops where braga was sold. At the same period, braga was consumed in large quantities by janissaries in the army due to its refreshing and beneficial properties on the body and a very low degree of alcohol (www.wikipedia.en).

In 1876 the brothers Haci Ibrahim and Haci Sadik opened a store in Vefa district of Istanbul where they sold braga, the most appreciated drink until nowadays. In Turkey today they sell a famous so-called "Vefa Bozacisi", braga Vefa, a sweeter and more flavorful drink with a creamy consistency (www.stiri-evenimente.ro). Today the company is managed by a great-great-grandson of Hagi Sadik and still produce one of the most popular varieties of braga, though, this drink no longer enjoys of the popularity in the past. However, the Vefa district of Istanbul has become a tourist attraction due to this special product made only from millet bark which is boiled with water, sometimes containing added sugar as well (www.wikipedia.en).

Nowadays in Romania, it is not easy to find braga, some decades ago it was popular in the Constanța area. Still, there are families that produce braga in Severin, Slatina and Giurgiu, and in Bucharest some restaurants have begun to produce and sell braga, a beverage that seems to be quite valuable, both in terms of nutrition and economical one (www.dambovita.net).

The only laboratory in Romania where braga is being made according to the traditional Turkish recipes, with manufacturing license and certificate attesting the product quality based on analyzes made by the Faculty of Food Industry within the "Lower Danube" University is the company SC COMALINA SRL. For producers, prospects regarding marketing of this product are not very encouraging, their output being adversely affected by the competition which produce counterfeit and poor quality braga.

2.2. The variety of the braga assortments

Braga is a traditional Turkish beverage with low alcohol content and is obtained after the fermentation process of yeast and lactic bacteria, produced by a mixture of cereals: millet, oats, barley, corn, wheat, rye or rice (Todorov S.D., Holzapfel W.H., 2014).

Although the traditional braga is obtained from millet flour and has the best taste, other variants were generated, containing a variety of grains or other ingredients being added, each recipe being improved by the specificity of each country. However, the original recipe is secret in order to avoid the use of fraudulent practices, this being already done by a number of contemporary manufacturers who want to bring the old taste of this beverage in an artificial and unhealthy manner. The true braga is a 100% natural fermentation beverage, obtained only from cereals with or without added sugar and yeast.

Depending on the region where this drink is produced, the main ingredients are millet in Turkey, wheat milled coarse in Kyrgyzstan, white wheat flour in Crimea, rice milled coarse in Turkmenistan and in Albania, the main ingredient is corn milled together wheat flour, sugar and water. In Serbia this drink is produced from wheat flour, corn flour, yeast, sugar and water, while in Romania braga is produced from a mixture of grains, from millet, barley, oats, wheat bran and corn flour. Other recipes mentions ingredients like barley, rye, chick and even rye bread (www.dambovita.net, www.121.ro).

Some specialists in this field consider that the braga which containing corn is a variant which deviate from the original recipe, but which still contains a variety of nutrients. Cornmeal was added by Armenians and thus the braga obtained in this manner has a very short shelf life, getting quickly a sour borscht taste (www.121.ro).

The authentic braga was different from the one that is being produced nowadays and has a high alcohol content (up to 7% by volume). In Egypt a traditional drink called "bouza" is still being produced. In the South African region, the production of braga became very important in the beverage industry and in other countries there is a variety of manufacture recipes and methods of this drink. In Turkey, the official name of the drink is "boza" and is consumed especially in winter but it is highly appreciated in summertime as well due to its refreshing character imprinted by the lactic acid. In summer, because of the high temperatures, changes in the sensory qualities and a reduction of the period of validity can occur (Todorov S.D., Holzapfel W.H., 2014).

Given the fact that the ingredients used are different from one country to another and the taste of the drink, how serving differ greatly, as well (www.wikipedia.en, www.dambovita.net, www.culinar.ro, www.tandfonline.com):

- in Turkey braga is creamy and sweet and is consumed more in winter, sprinkle with cinnamon, along with roasted chickpeas;

- in Bulgaria braga is very consistent and drunk often at breakfast together with a kind of pie called bushel ("banitsa with boza"). Although it is a popular drink, braga was associated with a lower quality product due to the color and cloudy appearance, so sometimes cocoa is add in. Until 1989 in Bulgaria braga was sweetened with sugar only, then they made appeal to the use of artificial sweeteners, aspartame being mostly used in order to increase preservation and cost reduction;
- in Kyrgyzstan, braga was sold by street vendors, especially in summer, being often mixed with milk like kefir which gives a prickly taste;
- in Albania, braga is weaker and is consumed mainly in summer. It is very popular and can be found in any store that sells ice cream or sweets, its taste varying from sweet to sour depending on the degree of fermentation;
- in Macedonia, braga has a much thinner consistency and tastes sweet. The most famous shop that sells braga is "Apche" – translated as "the pill". The name was chosen because people believed that braga is a cure for all diseases;
- in Bosnia and Herzegovina braga is considered an Eastern product sold in bakeries and nutritionists' shops, being made of a mixture of water, wheat flour, corn flour, yeast and sugar;
- in Egypt braga is characterized by a high content of alcohol (up to 7% by volume) being considered as beer;
- In Romania its specificity consists in its taste and appearance, being sweeter than in Turkey and Bulgaria, but denser and darker than in Macedonia. Here braga has several varieties, depending on the ingredients used, the technology being based on a mix of more grains whose composition differs depending on the area. This drink may or may not contain added sugar or honey. The liquid has a brown or beige-opaque color, and when hops are used the drink has a residual taste of beer.

By definition, braga is a refreshing sweet-sour drink, easily alcoholic, with a specific flavor due to fermentation of cereals used in its production.

The traditional recipe to obtain braga, produced currently in Turkey, specifies the following technology: preparation of raw material, boiling the cereal mixture, then cooling and filtration to remove solid particles, addition of sugar and proceeding of inoculation with a starter culture, represented by yogurt or sourdough, or in other cases, addition of yeast, followed by fermentation process conducted in wooden barrels. The braga obtained in a previous batch can be starter culture for another batch. The mixture is boiled at 30°C for 24 hours, cooled and kept in the refrigerator for 3-5 days (Hancioglu O., Karapinar M., 1997; Todorov S.D., Holzapfel W.H., 2014; Zorba M. et al., 2003).

Another recipe for this drink that preserves the traditional technology of Turkish fermented beverage envisages the following steps: the mixture of grains – millet, wheat, rye and barley, boiled for five hours only on wood fire, then cooling and letting to ferment for eight hours and then filtration. Honey or sugar and water are added, leave one day to cool to define the flavor and aroma, then it can be served. Braga becomes turbid, mucilaginous with very small amounts of alcohol and acid fermentation products (www.metropotam.ro).

The recipe which use only millet as a main ingredient, presents the following steps: the boiling of water with milled millet, and turbid liquid obtained after cooling, at the beginning of alcoholic and acid fermentation representing the fresh and sweet braga. Shortly after preparation, braga becomes a sour liquid, turbid, mucilaginous with small amounts of alcohol and lactic acid fermentation products (www.stiri-evenimente.ro).

Braga, also called the Sultan refreshing drink is prepared in the laboratory from Galați, according to an ancient recipe: cornmeal, whole wheat flour, yeast, sugar and water. The ingredients are mixed in certain proportions kept secret, after they are cooked in a special boiler for six or eight hours, depending on the quantity to be produced. To get quality braga, two things are important:
- the boiling cauldron, which is made from stainless steel, unsig only beech wood to boil il;
- the skill of the person who prepares the mixture of cornmeal and flour.

22

After allowing cooling for one day, it is filtered, and then the composition is mixed with water and the sugar or honey (www.adevarul.ro).

The recipe for braga "Alina" is based on thirty five years of experience of the company founders, Gheorghe Zmeu and Gabriel Pauliuc. Firstly, they learned the craftmanship from a Jew named Solomon Bal, who knew a traditional Turkish recipe, and secondly, they found out the secrets of this beverage inherited from the Turks, from their mother (www.producatorbraga.ro).

Another option for obtaining braga, used mostly in the Moldavia region and which represented the guide to produce samples of braga, proposes the following recipe: a mixture of equal parts (200g) of cereals: barley, rye, wheat bran, cornmeal and 600g millet grain. The grain is put in a pot with little water, enough to cover them and leave them to sprouting. The vessel can be covered with a towel and kept in a warm place to hasten the germination. After three days, water is drained off and the seeds are put to dry on plastic or enameled trays. After drying, they are ground with the meat grinder machine, resulting malt. The malt is mixed well with the cornmeal and wheat bran and put in a pot with boiling water, stirring continuously, for good uniformity. From this paste, cakes with a diameter of 10-15 cm are made, then these are put on a tray oven at high heat, until they get brown, then they are removed and left to cool. They are cut into small pieces and place in a pot (preferably wooden) with a capacity of 10-12 liters. Then a quantity of 6 liters of boiling water is poured over the cakes and left to ferment for 2-3 days. The liquid obtained with a brownish color, a specific smell and refreshing sweet-sour taste representing the braga. The same cakes can also be used again, but the second time smaller amount of water (approximately 4 - 4.5 liter) should be used, resulting lower braga (www.dambovita.net).

2.3. The characterization of cereals drink - braga

The braga virtues have been acknowledged since ancient times, being consumed with confidence by the Turkish army soldiers due to its refreshing properties, believing that it might render them more powerful.

It is supposed that braga was the starting point for making beer. Although the alcohol and acid content of braga was not known at that time, this drink was recognized as a stimulant and also as a medicine (Todorov S.D., Holzapfel W.H., 2015).

The physical and chemical properties of the beverage are significantly influenced by the raw materials, both during the fermentation and during storage period (Akpinar-Bayizit A., Yilmaz-Ersan, Ozcan t., 2010).

The differences which appear in the composition characterization of braga samples are due to the nature and quantity of cereal used as a raw material, some influencing a nutritional value of the product (Todorov S.D., Holzapfel W.H., 2015).

This ancient refreshing drink has a turbid aspect, mucilaginous, with colloid substance in suspension, due to its high content of extract nutrients, with very small amounts of alcohol and acid fermentation products (www.culinar.ro).

2.3.1. The nutritional valences of braga

The cereals which represent the raw material for the braga production have a high content of enzymes and phytonutrients, and they are found in the final composition of braga, and the taste and flavor resulting during the fermentation depend on the water used: the purer the water is, the more tasteful and better aroma it gets (www.bauturinaturale.ro).

Braga is a drink healthy and refreshing, containing carbohydrates, rich in vitamins A, B1, B2, B6, B12, C, niacin, biotin, folic acid, minerals, especially magnesium and potassium, lactic acid, enzyme, phytonutrients and many microorganisms and germs that can restore intestinal flora after antibiotic treatments (www.121.ro).

Depending on its acid contents, braga may present a sweet-sour taste, representing a nutritious beverage due to its high digestibility, its organoleptic characteristics imprinted by fermentation, but mostly to its content in nutrients: proteins, carbohydrates, fiber, vitamins and minerals and is a valuable food for human, being a source of nutrition for body's cells (LeBlanca J.G., Todorov S.D., 2011; Todorov S.D., Holzapfel W.H., 2014).

Braga contains about 0.5 – 1.61% proteins, 12.3% carbohydrates and 75 – 85% humidity (Yegin S., Uren A., 2008; Zorba et al., 2003).

The studies made on this drink showed that a cubic centimeter of sweet braga are 784500 germs, and in one cubic centimeter of sour braga, which is obtained from sweet braga after 5-8 days are 585000 of germs, thus being capable of restoring the saprophytic intestinal flora of the one who has used antibiotic treatment (www.bauturinaturale.ro).

The presence of various microflora in braga, in particular lactic acid bacteria and yeast, has a beneficial influence on the digestive system. This drink is recommended for athletes, pregnant women, vegetarians and especially for people suffering from intolerance to lactose.

2.3.2. Physicochemical characteristics of braga

In general, the pH of the braga samples varies between 3.16 – 4.02 and the average content of alcohol is 0.13% (m/V) (Köse E., Yücel U., 2003; Yegin S., Uren A., 2008).

Regarding the content of lactic acid, studies have shown that total titratable acidity was found to be the lowest in braga obtained from millet, ranging between 0.32 ± 0.04% and the highest one in wheat braga, 0.61 ± 0.07 %, this fact can be explained by the higher fermentable carbohydrates content of wheat compared to other grains. The pH of the beverage varies between 3.43 ± 0.08 to 3.86 ± 0.17 (Akpinar-Bayizit A., Yilmaz-Ersan, Ozcan T., 2010).

During storage, the acidity of the samples has increased (the highest recorded after 192 hours with 0.68 ± 0.06%) concomitantly with a decrease in the pH. In terms of alcohol content in wheat braga, a smaller amount, 0.46 ± 0.04%, was found, fluctuations during storage being observed, depending on the microbial and enzymatic activities. The acid and alcohol content depends mainly on the fermentation time, it was demonstrated that extended fermentation periods lead to an increase in acidity and the concentration of alcohol (Todorov S.D., Holzapfel W.H., 2014).

The preparing process of braga is characterized by two types of fermentation. Initially it was produced an alcoholic fermentation, with the production of carbon dioxide that contributes to increase volume, this type of fermentation quickly turns into lactic fermentation producing lactic acid and gives the acidic and refreshing character of the drink. Due to the fact that during fermentation an increase in volume is obtained, the wooden barrels must not be completely filled. Braga should be consumed within a few days since its production to avoid excessive fermentation which impart a sour taste. In practice, the duration of fermentation is retarded by cold storage of the product in order to extend the shelf life (LeBlanca J.G., Todorov S.D., 2011).

The starter cultures which are used to obtain the sourdough, can be yogurt or braga obtained in a previous batch. When using sourdough a less viscous product and lower acidity will result as compared to that obtained by yogurt inoculation, when a more viscous product with good acidity is obtained and yogurt taste is easily detected (Todorov S.D., Holzapfel W.H., 2014).

A study focusing on the sensory changes occurring in braga during the storage period has concluded that taste of braga samples was negatively influenced by developments in acidity due to the activity of lactic acid bacteria in the product, so the samples were considered by tasters generally acceptable for the first five days of storage, after which they showed a decrease in overall acceptability (Todorov S.D., Holzapfel W.H., 2014).

2.3.3. Microbiology of braga

Braga is a complex environment, characterized by a very diverse microbial population; the microorganisms responsible for flavor and pleasant taste are *Saccharomyces cerevisiae, Leuconostoc mesenteroides* and *Lactobacillus confusus* (Alan J. Marsh et al., 2014).

Numerous studies have been made in order to identify the lactic bacteria from braga using PCR and biomolecular techniques.

A study carried out on three traditional braga samples from Bulgaria which aimed at identifying microflora showed that this drink contains mainly lactic acid bacteria and yeasts. It was found that the number of bacteria was significantly higher as compared with yeast, highlighting five species of bacteria in different proportions, 82% belonging to the genus *Lactobacillus* and 18% belonging to the genus *Leuconostoc* five yeast species belonging to the genera *Saccharomyces* (47%), *Candida* (33%) and *Geotrichum* (20%) (Gotcheva V. et al., 2000).

Another study reported that the number of lactic acid bacteria identified in this drink ranged from 9×106 and 5×107 cfu/ml and was represented by *Lactobacillus paracasei subsp. paracasei, Lactobacillus pentosus, Lactobacillus plantarum, Lactobacillus brevis, Lactobacillus rhamnosus, Lactobacillus fermentum, Leuconostoc lactis* and *Enterococcus faecium* (Todorov S.D., Holzapfel W.H., 2014).

Braga contains a variety of lactic acid bacteria including strains of *Lactobacillus acidophilus, Lactobacillus brevis, Lactobacillus "coprophilus", Lactobacillus coryniformis, Lactobacillus fermentum, Lactobacillus paracasei, Lactobacillus plantarum, Lactobacillus rhamnosus, Lactobacillus sanfranciscensis, Lactococcus lactis subsp. lactis, Leuconostoc mesenteroides, Leuconostoc mesenteroide subsp. dextranicum, Leuconostoc raffinolactis, Leuconostoc lactis, Enterococcus faecium, Pediococcus pentosaceus, Leuconostoc oenos, Weissella confusa* and *Weissella paramesenteroides* (Arici M., Daglioglu O., 2002; Gotcheva V. et al., 2000; Todorov S.D., 2010; Todorov S.D., Dicks L.M.T. 2006; Von Mollendorff et al., 2006).

Other studies have sought isolation of yeasts and molds in braga, among yeast strains being found *Candida glabrata, Candida tropicalis, Geotrichum candidum, Geotrichum penicilatum, Saccharomyces carlsbergensis, Saccharomyces cerevisiae* and *Saccharomyces uvarum* (Arici M., Daglioglu O., 2002; Gotcheva V. et al., 2000).

2.3.4. Braga – a source of probiotics

Throughout history many fermented cereal products with multiple benefits for human nutrition have been produced, but the probiotic character of the microorganisms present in these beverages has been studied and reported later (Vasudha S., Mishra H. N., 2013).

In terms of twentieth century science, there were enough evidence to report the probiotic character properties owned by the lactic acid bacteria present in microbiota of this drink (Todorov S.D., Holzapfel W.H., 2015).

Due to its popularity several studies have been carried out on this drink, some describing lactic acid bacteria with probiotic properties of bacteriocin-producing.

Kollath defined the term of „probiotic" for the first time in 1953, including in this category of food probiotics all organic and inorganic complexes, characterizing them in antagonism with antibiotics that are harmful to intestinal microflora. Lilly and Stillwell proposed the following definition for the term of probiotics „microorganisms that accelerate the development of other microorganisms" (Vasudha S., Mishra H. N., 2013).

According with World Health Organization, probiotics are live microorganisms which when administered in adequate amounts have a beneficial effect on the host organism, reducing gastrointestinal infections and inflammatory intestinal diseases, with positive effects on the immune system (Todorov S.D. et al., 2008).

It was reported that probiotics play an important role in immunology, respiratory and digestive system and may have a significant effect in reducing the infectious diseases in children. Moreover, probiotic cultures improves the lactose digestion in people who suffer from lactose intolerance, increase the concentration of β-galactosidase, prevent allergies and reduce the risk of colorectal cancer and bladder cancer (Li J. et al., 2012, Rafter J., 2004).

It has been acknowledged that lactic acid bacteria produce antimicrobial compounds, including bacteriocins. A study carried in 2005 by Todorov and Dicks has reported that from a total of 52 isolates strains from braga produced in

Belogratchik, Northwest of Bulgaria, 13 strains showed growth inhibitory action against species *Pediococcus spp. Listeria innocua* and *Lactobacillus plantarum*. The total number of lactic acid bacteria identified in braga was 2×108 cfu/ml, revealing a antilisterial activity and lack of action against Gram-negative bacteria (Todorov S.D., Dicks L.M.T., 2006)

Other studies have shown that some strains, including species of *Lactobacillus plantarum, Lactobacillus pentosus, Lactobacillus rhamnosus* and *Lactobacillus paracasei* produced bacteriocins which act against pathogenic bacteria *Escherichia coli, Pseudomonas aeruginosa, Klebsiella pneumoniae Enterobacterium faecalis* and most bacteriocins show bactericidal action (Todorov S.D., Holzapfel W.H., 2014).

It was also found that two of the five isolates had antifungal activity with inhibitory activity against *Aspergillus niger, Penicillium spp Epicoccum* and *nigrum* (Todorov S.D., Holzapfel W.H., 2014).

Of all probiotics strains isolated from braga, ST284BZ strain (*Lactobacillus paracasei*) proved to have the best properties auto-aggregation with a high capacity of antibacterial and antiviral activity. Based on these characteristics, the strain ST284BZ is the best probiotic identified in braga, which may be used in cases of tuberculosis due to its ability to act destructively against the bacterium *Mycobacterium tuberculosis* (Todorov S.D. et al., 2008).

Lactic bacteria and, in small part, yeast produce vitamins and help increase the nutritional value of the product. Antimicrobial compounds, including bacteriocins, produced by lactic acid bacteria lead to the increase of product safety. In terms of safety, braga is a safe and healthy drink being recommended and promoted throughout society, especially in disadvantaged areas where there are people suffering from various diseases (http://nopr.niscair.res.in/bitstream).

2.3.5. Biogenic amines in braga

The biogenic amines are found in many foods especially in products that require a maturing or fermentation period, such as cheese, meat products, beer, wine and fish. Food consumption containing biogenic amines can cause metabolic disorders and various disease such as headaches, nausea, vomiting, hypotension, hypertension, heart palpitations. Fermented foods and beverages are less sensitive to the biogenic amines formation because most strains are involved in the fermentation process and the raw materials used contain considerable amounts of proteins.

Although there are not many studies on the possibility of the formation of biogenic amines in braga, some researchers have shown their interest in this issue given the fact that the raw materials used are a rich source of proteins constituting a substrate for a part of microorganisms with decarboxylase activity (LeBlanca J.G., Todorov S.D., 2011).

After analyzing the 10 samples of braga purchased from different manufacturers in Turkey, using HPLC technique after derivation with benzoyl chloride it can be noticed that *Lactobacillus sanfrancisco, Leuconostoc oenos (Oenococcus oenus)* and *Leuconostoc mesenteroides* are biogenic amine-producing bacteria. Of the 11 biogenic amines, three of them were found in all samples: putrescine, spermidine and tyramine, the tyramine being found in the highest concentrations, ranging between 13 and 65 mg/kg. The total content of biogenic amines in braga samples was between 25 and 69 mg/kg (Yegin S., Uren A., 2008).

In another study which aimed to identify biogenic amines in braga, 21 samples were analyzed and the presence of six biogenic amines was identified: histamine, tryptamine, beta-phenylethylamine, putrescine, cadaverine and histamine. Out of 21 samples, in 18 samples at least one of the six biogenic amines listed was identified. The total of biogenic amines ranged from 1.67 to 101, 14 mg/kg, the highest amount being tyramine or 82.79 mg/kg, followed by cadaverine, tryptamine, putrescine, beta-phenylethylamine and histamine respectively 17,69; 13,78; 9,80; 4,53 and 4,07 (Cosansu S., 2009).

Although the content of biogenic amines resulted from the analysis is below the toxic limits recommended, the braga consumption might constitute a health risk, especially in patients treated with drugs containing monoamine oxidase inhibitor.

2.4. The Galați braga – certificated product in Romania

Motivated by the taste and appreciation of old cereal beverage - braga, two Galatians, Gheorghe Zmeu and Gabriel Pauliuc administrators of SC COMALINA SRL, have developed a business that aims to promote, manufacture and market braga. Although in the country there are areas where braga is still being produced, the two men own the only laboratory where braga is being prepared with manufacturing license and certificate attesting of product quality based on the analyzes carried out at the Faculty of Food Industry within the „Lower Danube" University.

The braga product obtained at the SC COMALINA SRL, is produced in two varieties, braga and cream braga by boiling the mixture of whole wheat flour, cornmeal, sugar, with or without honey or artificial honey, yeast and water.

According to the technical specifications, the ingredients used in the preparation of the two products are shown in table 2.1.:

Tabelul 2.1.

The ingredients used in the manufacture of braga

(technical specification SC Comalina SRL)

Product name	Raw material	U.M.	Quantity
Braga	whole wheat flour	kg	9
	cornmeal	kg	9
	granulated sugar	kg	4,8
	artificial honey	kg	6,5
	yeast	kg	0,07
	whater	l	72,0
Cream braga	whole wheat flour	kg	9
	cornmeal	kg	9
	granulated sugar	kg	4,8
	artificial honey	kg	6,5
	yeast	kg	0,07
	salt	kg	0,01
	whater	l	72,0

In order to obtain a properly product in terms of nutritionally and hygiene aspect, its it aimed that the raw and auxiliary materials should comply with all conditions of technical quality, which must correspond to the normative documents of product and sanitary rules in force. In this sense quantitative and qualitative analyze of the raw materials in terms of organoleptic, physical-chemical and microbiological views should be made.

The braga product is manufactured in conformity with the technological instructions, the recipe and the technological phases of manufacturing scheme approved by the company's management in compliance with the sanitary rules in force. The technology of braga product represent a set of technological operations, after which the raw and auxiliary materials used in this process are converted into a finished product. The main operations of manufacturing process of this beverage are shown in figure 2.1.

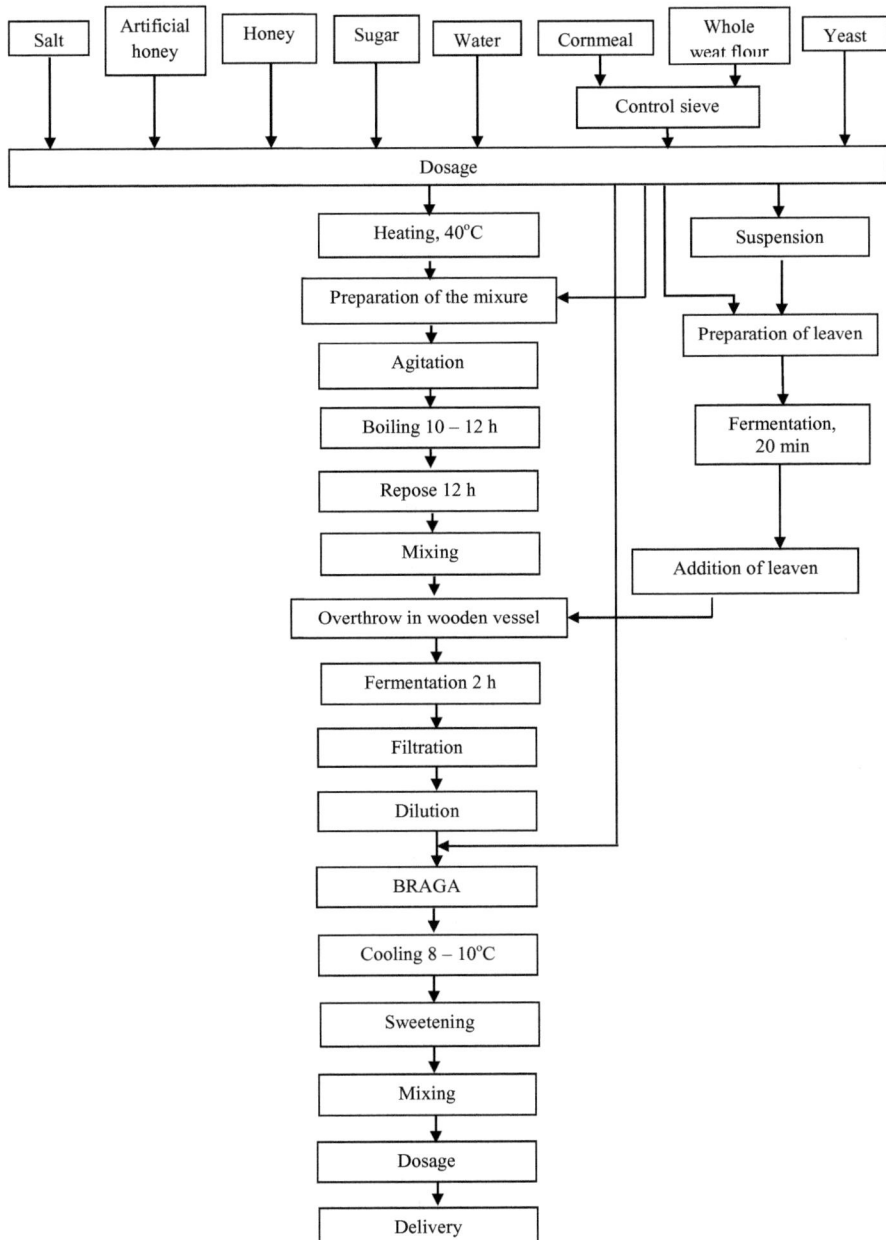

Figure 2.1. The flow diagram of braga Alina
(technical specification SC Comalina SRL)

The two varieties are produced by boiling a mixture prepared from wheat flour, cornmeal and water. The mixture resulting was cooled, then yeast is added and allowed to ferment in wood vessels. Finally, sugar and honey or artificial honey are added, and then the mixture is poured into clean and dried dishes and is delivered. Figure 2.2. shows the flow technology of braga beverage from SC COMALINA SRL Galați:

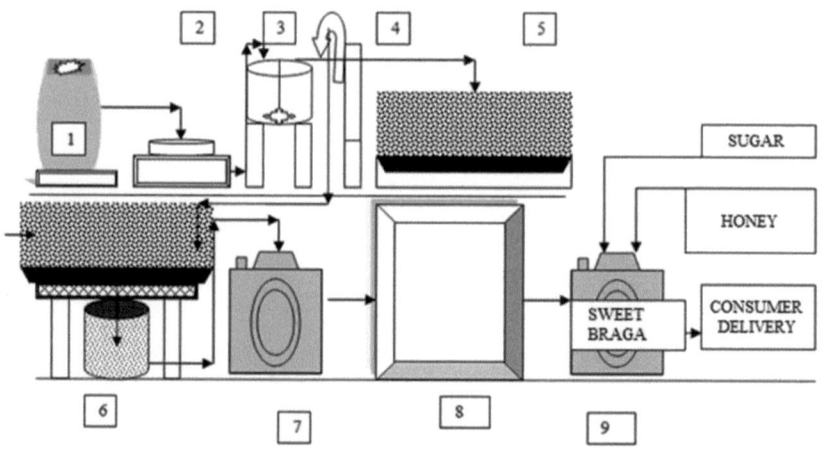

Figure 2.2. The technological sketch of manufacturing flow in S.C. Comalina S.R.L.
(1 – weighing raw materials, 2 – control sieve, 3 – boiler for boilling mixture, 4 – water and raw materials dosage, 5 – fermentation vat, 6 – sieve system for straining the mixture of braga, 7 – canisters filled with braga, 8 – cold room,
9 –sugar or honey addition)
(technical specification SC Comalina SRL)

The product is stored only in specially designated areas. They must be clean, bright, airy, free from mold, insects or rodents, isolated from strong heating source, to a uniform room temperature 2 – 4°C. The drink is stored unsweetened, the sweetening being made upon delivery to the consumer.

The shelf life of the product is of 48 hours, established since the product is obtained in the production department and refers to the product's maintenance under appropriate conditions of humidity and air temperature, in proper hygienic conditions (temperature 2 – 4°C and relative air humidity 70 - 80%). The unit subdivision of SC COMALINA SRL, according to the company's technical specification is shown in figure 2.3.

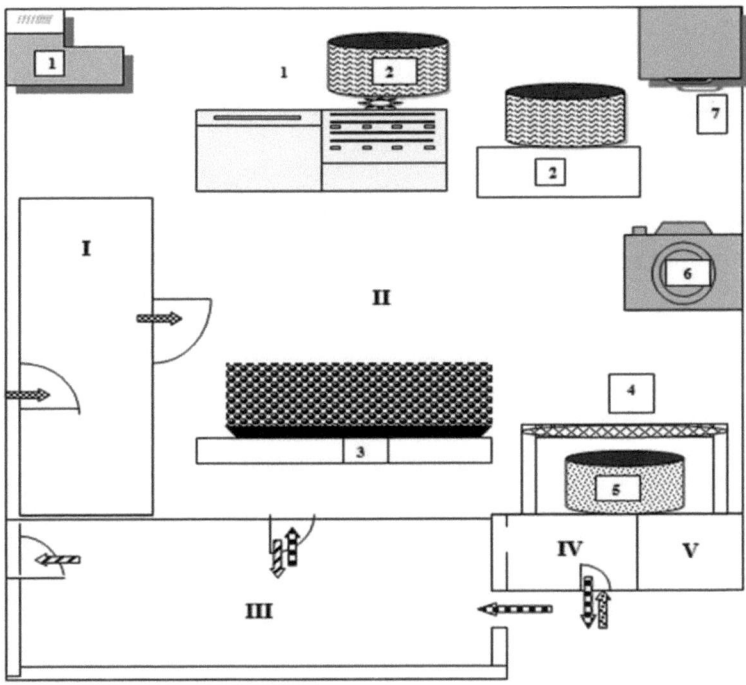

Figure 2.3. The unit subdivision SC Comalina SRL

(I – storehouse of raw materials, II – room manufacturing braga, III – hall, IV – staff cloakroom, V –sanitary space, 1 – weighing for raw materials, 2 – boiler for boillingthe the raw materials, 3 – vat for the mixture fermentation, 4 – sieve for straining the braga, 5 – braga colecting vessel, 6 – canisters with braga, 7 – fridger)

Staff circuit	
Laborers' circuit	
Raw materials circuit	
Finished product circuit	

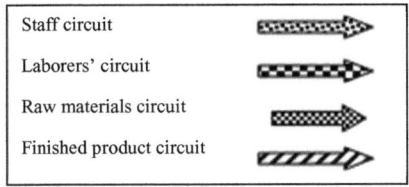

(technical specification SC Comalina SRL)

The quality of the finished product obtained by SC COMALINA SRL is verified on each lot and periodically, both in the shop floor and within the laboratory. Forevery batch packaging and marking, the organoleptic properties, physical and chemical properties and the microbiological examination are being checked.

The checking of the packaging and labeling is made by examining the general batch how braga is being stored in throughout distribution and dispaly.

The verifying of organoleptic, physico-chemical and microbiological properties is carried out on a representative sample which is properly collected, considering that these samples corresponds with the quality standards requested.

The finished product is a drink obtained from boiling whole grain flours (wheat and corn), resulting a smooth paste, which due to fermentation becomes a yellowish, turbid, creamy, acidified liquid with colloid substance in suspension. According to the technical specifications, braga „Alina" produced by SC COMALINA SRL has the following organoleptic and physico-chemical characteristics:

Tabel 2.2.

Organoleptic characteristics of braga Alina

(technical specifications SC Comalina SRL)

Characteristics	Admissibility conditions	
	Braga	Cream braga
General aspect	turbid liquid, with milky aspect	consistent creamy liquid
Consistence	liquid with suspension	fluid, slightly creamy
Color	yellowish to dark beige	yellowish, dark beige
Flavor	specific fermentation	specific fermentation
Taste	acidulous, pungent, slightly sweet	acidulous, pungent, sweet
Smell	pleasant, typical of fermented	pleasant, typical of fermented

Physico-chemical properties of braga Alina

(technical specificationsSC Comalina SRL)

Variety	Water (%)	S.U. (refractometric degrees, minimum) / of which carbohydrates:	Total acidity, (lactic acid %) maximum	Ethyl alcohol (%) maximum	Volatile acidity (%), maximum	Carbon dioxide (%), minimum
Braga	83	7 / 14	0,4	1	0,04	0,1
Cream braga	80	8 / 15	0,4	1	0,04	0,1

Braga meant for human consumption must not contain the altered raw materials, synthetic sweeteners, emulsifiers or preservatives. In case of nonconformities appropriate measures for quality assurance should be taken.

CHAPTER 3

DESCRIPTION OF AROMATIC PROFILE OF CEREAL
BEVERAGE - BRAGA

The aroma is one of the most important characteristics of a food product, particularly a fermented beverage and is defined by a number of volatile compounds that can stimulate the olfactory receptors.

The volatile compounds that contribute to the flavor of a food product have different origins: specific to the raw material used, occur during the process of fermentation, or are defined at the end of fermentation and belong to a large variety of substances: terpenes, alcohols, esters, volatile fatty acids, phenols and other (José M. Oliveira et al., 2009).

From a quantitative perspective these compounds are found in a small percentage in the product; they are separated and determined with difficult, but there is a high chemical diversity, especially in food products which have been subjected to the fermentation or thermal processes. Studies have shown that, of these compounds, less than 5% contribute to the flavor of the product and are characterized as impact compounds (Iordache A.M., 2011; Maarse H et al, 1992).

Despite of the strong smell, braga is highly appreciated due to its refreshing and energizing drink, especially for its healthy beneficial on the body.

The specific aromas that imprint braga particular flavor characteristics, a large part of them originated in the malt used for its preparation, after the fermentative processes and after the transformations occurring during manufacture, their concentration being influenced by the conditions in which the product undergoes during fermentation. Of the volatile compounds defining the flavor, we mention esters, alcohols, aldehydes, ketones, volatile phenols, lactones, free fatty acids and other various chemicals.

3.1. Materials and methods

Of all the methods of separating different substances from a mixture, used in analytical chemistry and technology, the chromatographic method is considered the most efficient. Chromatography is a method of separation of multi-component mixtures which is based on the different distribution of a mixture's components between a mobile phase and a stationary phase, being in a relation of a relative motion to each other (the mobile phase and a stationary phase), resulting the movement with different speeds of the components carried by the mobile phase along the stationary phase. The difference between the speeds' migration of components is specific to chemical nature and is dependent on the physico-chemical properties. Staggered over time, the components are worn from eluent solution, after leaving the column into the detector. This transforms the difference of the physical properties between component and eluent to an electrical signal that is proportional with the concentration of component in the gas mobile phase. The graphical representation of the detector signal depending on time is constituted in a chromatogram. The chromatographic separation is the result of repeated processes sorption – desorption of the sample components between the stationary phase and mobile phase (Gutt Sonia, Gutt G., 2005; Soceanu Alina Daria, 2009).

Gas chromatography include all chromatographic methods in which the mobile phase is gaseous, the technique allows the mixtures' separation of gases or liquids that can be volatilized at not very high temperatures. The separation aims at the different distribution of gaseous components between a mobile phase and a stationary phase. The main components of a gas chromatograph are: the carrier gas tank which acts as a mobile phase, the injection device for sample introduction, the column and the register (Soceanu Alina Daria, 2009).

In practice, various techniques are being used that suppose the combination of chromatographic separation system with one or more of the spectral analysis system in order to obtain more efficient analytical methods. The most common ones are those made by coupling chromatographic techniques (characterized by selectivity and efficiency of separation) with mass spectrometry (which provides structural information and increased selectivity) (Soceanu Alina Daria, 2009).

The coupling of the two methods is very advantageous, allowing the chromatographic separation of mixture components, which is then analyzed using a mass spectrometer. It is the most efficient method of separating components of a mixture, using very small amounts of sample, the only disadvantages being that it does not allow direct identification of components.

The technique combining gas-chromatography with mass spectrometry (GC-MS) is very useful for the identification and the study of volatile compounds present in various alcoholic and non-alcoholic beverages, in very small quantities. Following the analysis a specific chromatogram, considered the footprint of bouquet beverage, is being generated (Iordache A.M., 2011).

In the research of braga a gas-chromatograph coupled with mass spectrometry GC-MS QP2010 Plus, Shimadzu within Instrumental Analysis Laboratory of the Faculty of Food Engineering, Suceava was used (figure 3.1.). The flavor compounds were separated using a capillary chromatographic columns CP-sil88 with the following dimensions: length – 50 m, outer diameter - 0.32 mm, internal diameter - 0.2 µm. Stationary phase is siloxane, the flow rate of helium - 1 ml/min. Gas chromatograph is connected to a computer with program for the data registration, data processing is made using soft GC-MS Postrun Analysis. The compounds are identified using the database of mass spectrometry NIST.

Figure. 3.1. Gas-chromatograph coupled with mass spectrometry GC-MS QP2010 Shimadzu

In order to identify volatile compounds in braga via gas-chromatographic technique combined with the mass spectrometry, the sample was processed as follows: a mixture of 5 ml of the sample with 2 g of NaCl was introduced in a glass bottle covered with a rubber stopper and was thermostated for one hour at a temperature of 85°C in order to separate volatile compounds. Using a syringe the resulting vapors were extracted and injected in the chromatography column after the working specific parameters of computer program had been specified. The duration of the analysis was of 50 minutes, while the computer monitor generated a specific chromatogram, its interpretation being achieved by means of software GC-MS Postrun Analysis provided by the computer software.

3.2. Result and Discussion

Having as main objective the aromatic profile of natural drink braga, by carrying out gas-chromatographic analysis a number of 64 compounds was identified. Only some of these compounds are responsible for flavor, highlighting the mainly esters.

Figure 3.2. (a and b) shows the chromatogram generated after the analysis and the identification of compounds, red points representing each of the components identified, including those who belong to the column. The highest peaks are specific compounds in the column, such as siloxane and naphthalene.

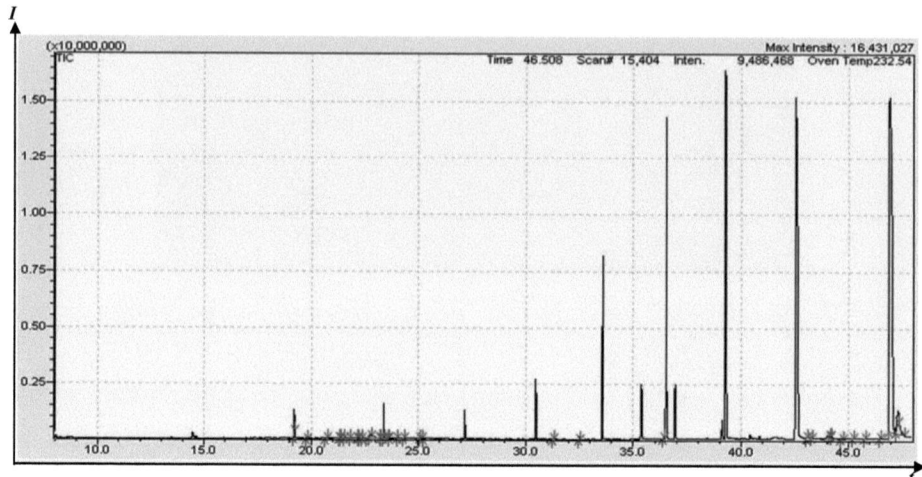

Figure 3.2.a. Gas chromatogram of volatile compounds in braga (during identification of compounds)

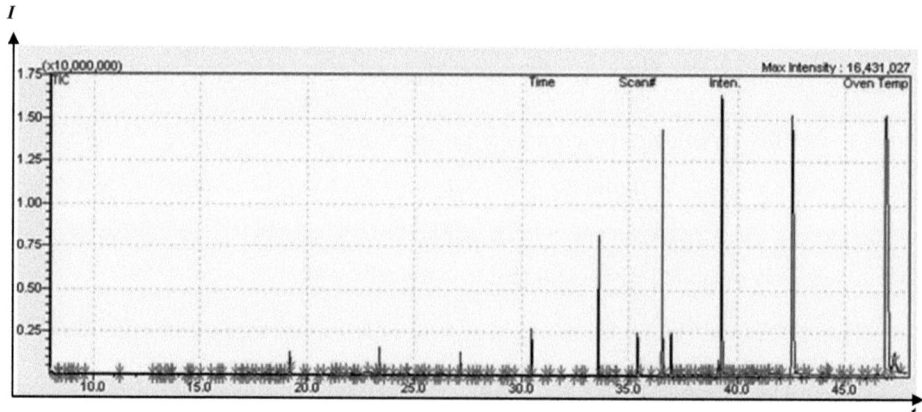

Fig.3.2.b. Gas chromatogram of volatile compounds in braga (after identification of compounds)

Among compounds analyzed: 20 esters, 19 alkanes, 12 acids, 5 amines, 4 alcohols, 2 aldehydes, ketones and other four organic compounds containing chlorine and iodine were identified.

The name of compounds identified in braga

Name of compound	Chemical formula	Time (minutes)
Pentafluoropropionic acid, undecyl ester	$C_{14}H_{23}F_5O_2$	8.6
Trichloroacetic acid, undecyl ester	$C_{13}H_{23}C_{13}O_2$	13.1
Decane, 1-iodo-	$C_{10}H_{21}I$	13.3
Eicosane	$C_{20}H_{42}$	13,4
Dodecane, 2,6,11-trimethyl-	$C_{15}H_{32}$	14.4
2-Propyl-1-heptanol	$C_{10}H_{22}O$	15.8
Hexadecane	$C_{16}H_{34}$	15.9
4-Methyldodecane	$C_{13}H_{28}$	16.7
1-Dodecanamine, N,N-dimethyl-	$C_{14}H_{31}N$	17.35
5,9-Dimethyl-1-decanol	$C_{12}H_{26}O$	17.6
Eicosane	$C_{20}H_{42}$	17.9
Eicosane	$C_{20}H_{42}$	18.1
Dodecane, 2-methyl-	$C_{13}H_{28}$	18.7
Tetracosane	$C_{24}H_{50}$	18.8
Octadecane	$C_{18}H_{38}$	19.8
Hexacontane	$C_{60}H_{122}$	20
2-Octanone	$C_8H_{16}O$	20.7
Trichloroacetic acid, undecyl ester	$C_{13}H_{23}Cl_3O_2$	21.3
Octadecanoic acid, 2,3-bis[(1-oxotetradecyl)oxy]propyl ester	$C_{49}H_{94}O_6$	21.4
Methoxyacetic acid, 4-hexadecyl ester	$C_{19}H_{38}O_3$	21.55
4-Fluoro-1-methyl-5-carboxylic acid, ethyl(ester)	$C_7H_9FN_2O_2$	22.9
Methoxyacetic acid, 4-hexadecyl ester	$C_{19}H_{38}O_3$	23.15
1-Chlorodocosane	$C_{22}H_{45}Cl$	23.6
Triethylene glycol monododecyl ether	$C_{18}H_{38}O_4$	24
Formic acid, decyl ester	$C_{11}H_{22}O_2$	24.6
2-Dodecenol	$C_{12}H_{24}O$	24.7
Propanoic acid, 2-methyl-, 2,2-dimethyl-1-(2-hydroxy-1-methylethyl)propyl ester	$C_{12}H_{24}O_3$	25.5
Propanoic acid, 2-methyl-, 1-(1,1-dimethylethyl)-2-methyl-1,3-propanediyl ester	$C_{16}H_{30}O_4$	25.85
Propanoic acid, 2-methyl-, 2-ethyl-3-hydroxyhexyl ester	$C_{12}H_{24}O_3$	26.7
1-Hexacosene	$C_{26}H_{52}$	26.1
1-Chloroeicosane	$C_{20}H_{41}Cl$	26.3
Tricyclo[20.8.0.0(7,16)]triacontane, 1(22),7(16)-diepoxy	$C_{30}H_{52}O_2$	27
4-Methylhexadecane	$C_{17}H_{36}$	27.7
1,3-Pentadiene, 2,4-di-t-butyl-	$C_{13}H_{24}$	28.4
Hexadecanoic acid, 1-methylethyl ester	$C_{19}H_{38}O_2$	28.6
Tetrapentacontane	$C_{54}H_{110}$	29.1
Benzoic acid, 2-ethylhexyl ester	$C_{15}H_{22}O_2$	29.2
Acetic acid, 3,3,6-trimethyl-4-oxo-3,4-dihydro-2H-pyran-2-yl ester	$C_{10}H_{14}O_4$	31.25
Cyclopentaneacetic acid, 3-oxo-2-pentyl-, methyl ester	$C_{13}H_{22}O_3$	34.1
Hexanedioic acid, mono(2-ethylhexyl)ester	$C_{14}H_{26}O_4$	34.35
Caprolactam	$C_6H_{11}NO$	35
3,5-di-tert-Butyl-4-hydroxybenzaldehyde	$C_{15}H_{22}O_2$	36.4
1(3H)-Isobenzofuranone	$C_8H_6O_2$	37.6

1,2-Benzenedicarboxylic acid, butyl 8-methylnonyl ester	$C_{22}H_{34}O_4$	38
Diphenylamine	$C_{12}H_{11}N$	38.4
Cyclohexanecarboxylic acid, dodecyl ester	$C_{19}H_{36}O_2$	38.6
Hexanamide	$C_6H_{13}NO$	39.6
Cholest-5-en-3-ol (3.beta.)-, nonanoate	$C_{36}H_{62}O_2$	40.7
l-(+)-Ascorbic acid 2,6-dihexadecanoate	$C_{38}H_{68}O_8$	41.6
9-Hexadecenoic acid, octadecyl ester	$C_{34}H_{66}O_2$	43.3
1,2-Benzenedicarboxylic acid, mono(2-ethylhexyl) ester	$C_{16}H_{22}O_4$	44.25
2-Propenoic acid, 3-(4-methoxyphenyl)-, 2-ethylhexyl ester	$C_{18}H_{26}O_3$	44.7
Tetradecanoic acid, tetradecyl ester	$C_{28}H_{56}O_2$	46.1
9-Hexadecenoic acid, octadecyl ester	$C_{34}H_{66}O_2$	46.5

Of these, the compounds conffering specific flavor of braga are esters, for example pentaphluoropropylundecyl ester; methoxiacetyl,4-hexadecyl ester; propionyl,2-methyl,2,2-dimethylpropyl ester; octylbenzyl ester; trimethyl-4-oxo,3,4-hydroxy-2-piran ester; benzendicarboxylmono ester; some of the acids, such as trichloroacetic acid and ascorbic acid (figure 3.3.i.) which imprint a sweet-sour taste, and some of alkanes, such as hexadecane, octadecane, eicosane.

Details of gas-chromatogram specific to braga are shown in figures 3.3.a. – 3.3.j. The generating of specific peaks of volatile compounds was achieved after 8 minutes since the sample injection.

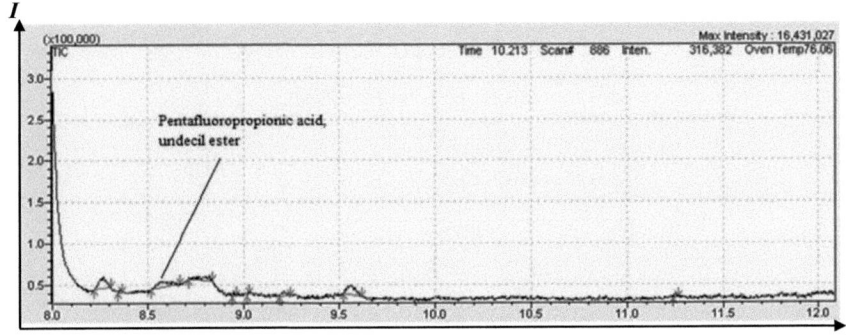

Figure 3.3.a. Detail of Figure 1 for the first 12 minutes

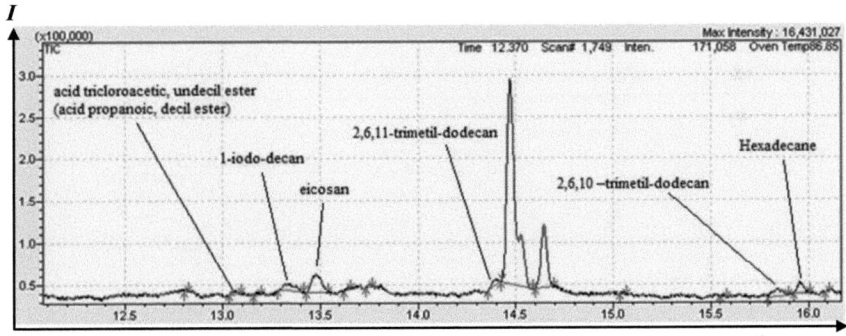

Figure 3.3.b. Detail of Figure 1 for the minutes 12-16,2

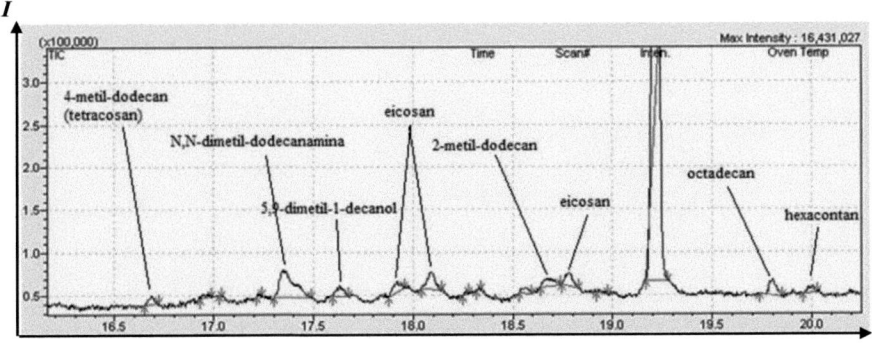

Figure 3.3.c. Detail of Figure 1 for the minutes 16.2 -20,2

45

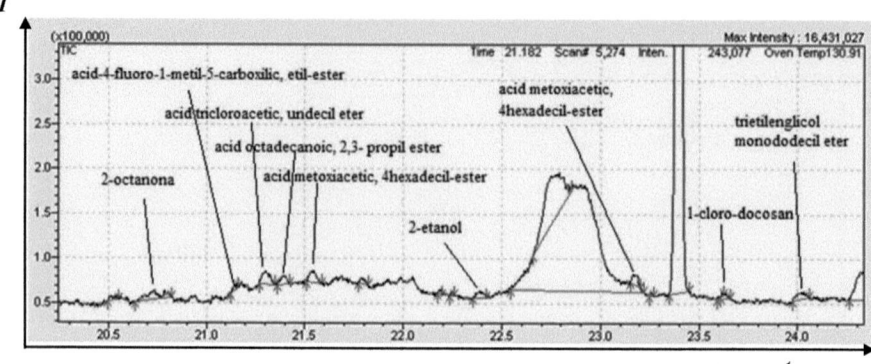

Figure 3.3.d. Detail of Figure 1 for the minutes 20.3 -24,3

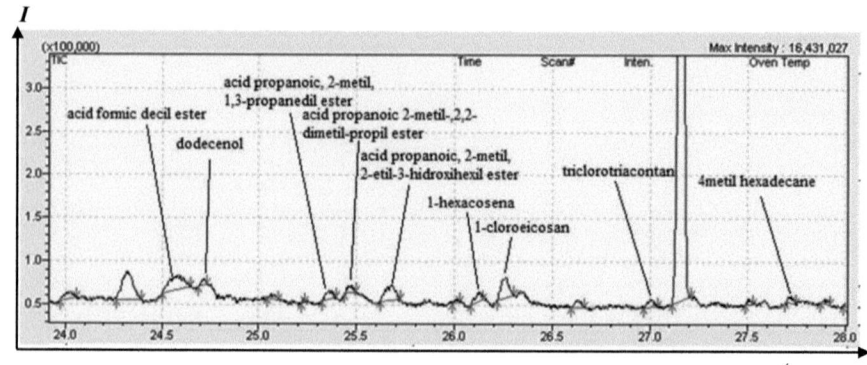

Figure 3.3.e. Detail of Figure 1 for the minutes 24.4 -28,0

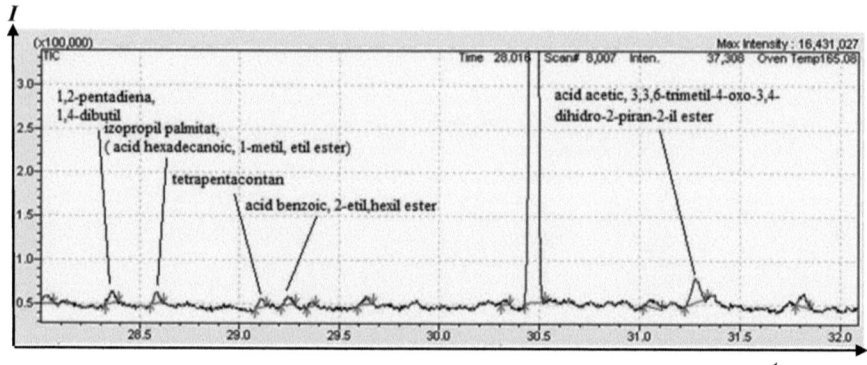

Figure 3.3.f. Detail of Figure 1 for the minutes 28.1 -32,0

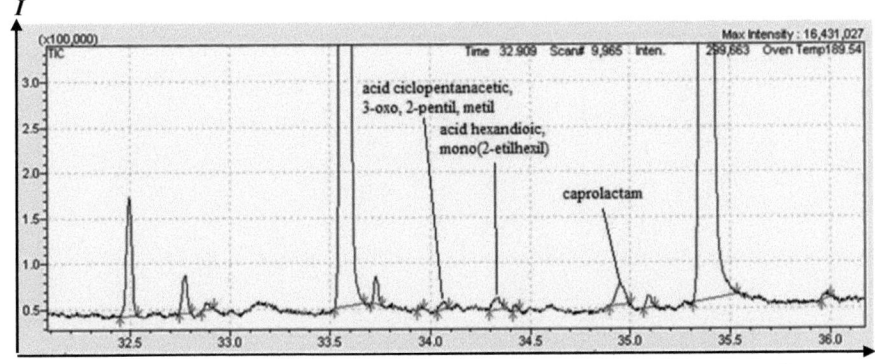

Figure 3.3.g. Detail of Figure 1 for the minutes 32,1 -36,2

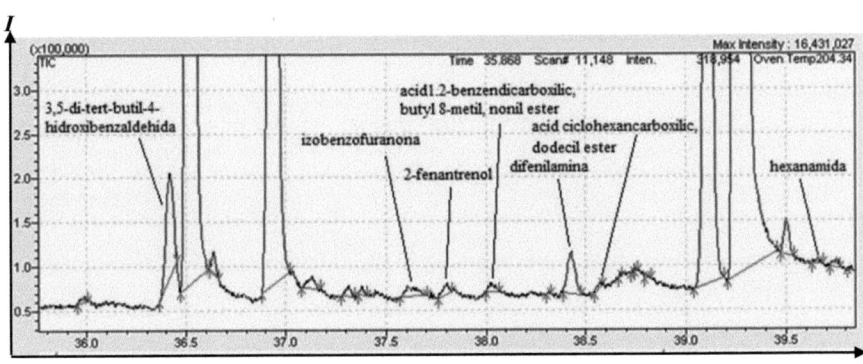

Figure 3.3.h. Detail of Figure 1 for the minutes 36.3 -40

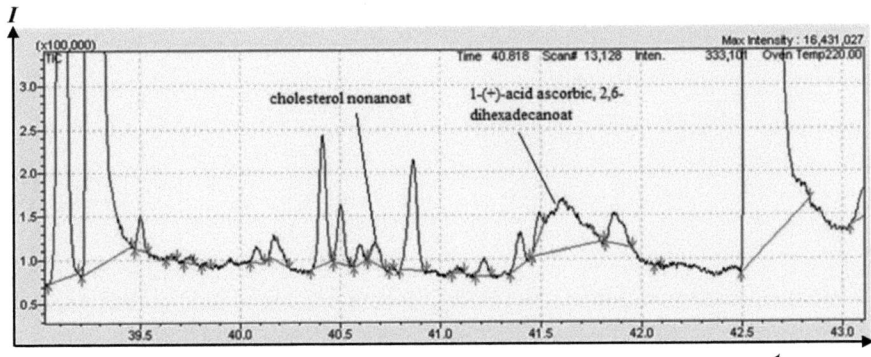

Figure 3.3.i. Detail of Figure 1 for the minutes 40,1-43.0

I

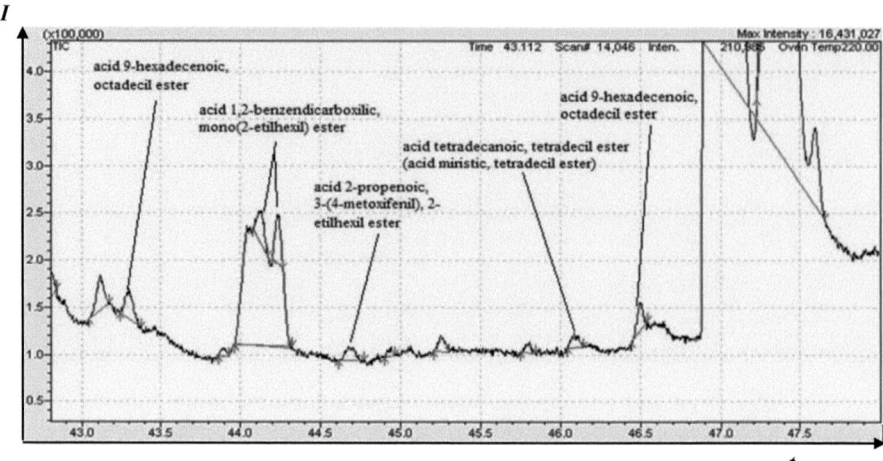

Figure 3.3.j. Detail of Figure 1 for the minutes remain -50

Following the identification of the flavor compounds, the chromatogram recorded as well other volatile compounds that did not imprint special properties to the product, such as phthalates (diethylphthalate, dibutylphthalate) or amines (diphenylamine, caprolactam), whose source is the raw materiasl - cereals, the soil in which they were grown or the most likely the treatments applied in agriculture (pesticides, insecticides).

CHAPTER 4

CHARACTERIZATION OF CEREAL BEVERAGE - BRAGA IN TERMS OF TRACE ELEMENTS AND HEAVY METALS

Food represents the biological relationship between man and environment and ensures the basic nutrient substances and energy for a good development of metabolic processes, growth and development of the individual.

The fast pace of modern lifestyles has led to changes in the individual's concern for an appropriate diet. Confused by the phenomenon of "food's explosion" manifested through the extreme diversification of food categories and the complexity of the chemical composition thereof, the consumer is increasingly interested if it complies with the standards in terms of quality and food safety, aiming to ensuring the products' quality and the reduce risk of contamination.

Foods can be influenced and affected by a variety of internal and external factors, both at the site of its production, and during the technological process or its handling. Therefore in the last period, mare and more attention was given to the nutritional value and innocuousness status of the product.

Food quality and safety has in view the effort of all those involved in the complex chain of producing it - agricultural production, processing, transportation and consumption, all these becoming a responsibility of every individual, beginning with the products' origin until the moment when they are ready for consumption. So, those directly or indirectly involved in the food chain have the responsibility to provide safe products for consumption in line with consumers' final expectations (Cara Daniela, Vlad Gh., 2014).

Food quality and safety is an intrinsic characteristic, this issue being considered a right of the consumer to use the safe product for consumption, with direct effects on their life, ensuring their safety and protection their interests.

According to the Codex Alimentarius, food safety is "the guarantee that food will not cause harm to consumers when they are prepared and/or eaten according to the specifications on how to use".

In order to maintain the food quality and safety, quality procedures are required to ensure that foods are intact, and monitoring procedures as well to be sure that the technological processes are operated in appropriate conditions. To ensure that marketed food is safe and healthy a number of norms, standards, codes of good practice, international and domestic hygiene codes that must be met throughout the food chain were developed (Cara Daniela, Vlad Gh., 2014).

This can prevent, or at least identify any source that would cause contamination of the food. The most common statements of illness (food poisoning) caused by food are of microbiological origin, given that microorganisms are everywhere and can enter at any point in the food chain. Therefore, lately there have been more and more concerns with the consequences of chemical contamination, because of the "modernization" in all segments of production. The environmental pollution by industrial activities, transport or the relaxation activities of the individual ultimately lead to pollution of the organism, with direct implications for physiological and psychological state of the individual.

Vegetables organism have a complex chemical composition and therefore is not very easy to identify chemical compounds that are part of them because their structure differ both by nature and proportion of components and through the arrangement of atoms in the molecule (Soceanu Alina Daria, 2009).

Having in view that living matter is composed of macro and micronutrients whose amount varies from the order of grams parts per million (ppm) or billion (ppb), it has been demonstrated that some of these are absolutely necessary, for example, Mg, Na, Co, Ni, Cu, Zn, Cr, Mo, Mn, Se, while other are essential such as Al, Ba, Sr, Rb, B, Li, and some non-essential or tolerable within certain limits: As, Hg, Cd, Pb, Au, Ag, U (Iordache A.M., 2011).

The nutrients contained in living matter play an important role in the activation of some enzymes, in the synthesis of some intermediates or various metabolic processes. Thus, Co, Fe, Mg, Mn, Zn, K plays an important role in the metabolism of carbohydrates as activator of phosphatides, and Ca, Mn, Mg, Zn, Fe, and Ni involved in the synthesis of peptides. Elements such as Mn, Mg, Zn, Fe, Co, K, Rb and ammonium interfere with the activation of enzymes such as aconitase, dehydrogenase and decarboxylase, which are involved in the respiratory or synthesis processes (Soceanu Alina Daria, 2009).

It is essential to understand the need for optimum ratios of these elements because the absence of one or presence of another elements in too high concentration lead to metabolic imbalance, illness or poisoning, as it is the case of heavy metals.

Along with minerals, considered responsible for the beneficial actions on the body, braga may contain less desirable micronutrients, such as heavy metals coming from the raw materials used in manufacture or it may not comply with the health and hygiene conditions. These elements can contaminate the product during the technological process.

It is desirable that any food, especially those which have special nutritional properties, to be free from any source that could jeopardize consumers' health and comply with applicable regulations and standards regarding food quality and safety.

4.1. Materials and methods

Mass spectroscopy is the most sensitive method of structural analysis, using a micro-analytical technique that allows to measurement of relative molecular masses of unitary compounds, as well as highlighting certain atomics and functional species existing in the compound analyzed. This analytical method allows ions' separation on the basis of the mass-charge ratio (m/z) and ions' registration, using quantitative and structural determination of atoms and molecules (Culea M., 2008; Gutt Sonia, Gutt G., 2005).

Mass spectroscopy allows the development of methods to determine the content of heavy metals in drinking water and beverages, using inductively coupled plasma mass spectrometry (ICP-MS), technique used for the quantitative multi-element determination, at the level of traces (mg/l). The method is used for the analysis of increased radioactive samples by the fact that in the part of optical analysis does not enter radioactivity, and the plasma gases are absorbed 100%. The great advantage of plasma spectroscopy is that due to the high temperature of plasma, it is the most efficient method for multi-element spectroscopic analysis with the possibility of simultaneous analysis of 70 elements in a sample. Basically, a solution can be analyzed in a minute allowing analysis of large numbers of samples in a short time at low detection limits down to the level of parts per trillion (ppt) (Gutt Sonia, Gutt G., 2005; Iordache A.M., 2011).

The technical of inductively coupled plasma mass spectrometry (ICP-MS) was introduced in 1983 and soon gained acceptance to a large number of laboratories due to its detection capabilities superior to other techniques.

The inductively coupled plasma mass spectrometry use the samples transformed into an aerosol or placed in the plasma. Initially the samples are subjected to a process of preparation which consists of filtration, acid digestion and dilution. Depending on the accuracy of this stage, the quality of final results is generated, since a larger quantity of the sample leads to a stronger signal to the detector, better statistic, low detection limits and high quality results (Claudiu Tănăselia, 2013).

The thermic plasma is an ionized gas, of high temperature to 10000K which in addition to carbon atoms, contains electrons and ions (Gutt Sonia, Gutt Gh., 2003).

The plasma – mass spectrophotometer interface is a constructive unit of the system through which the ions formed in plasma are taken up and transported into a fascicle form to detector (Tănăselia C., 2013).

The research concerning the analysis of heavy metals and trace elements in braga used a mass spectrometer with inductively coupled plasma ISP-MS Agilent 7500, within Instrumental Analysis Laboratory of the Faculty of Food Engineering, Suceava (figure 4.1.).

The parameters of ICP-MS have been: nebulizers - 0.9 ml/min, RF power 1500W, carrier gas – argon (0.92L/min), the mass-range 7-205, integration time 0.1 seconds, the acquisition 22.7 seconds.

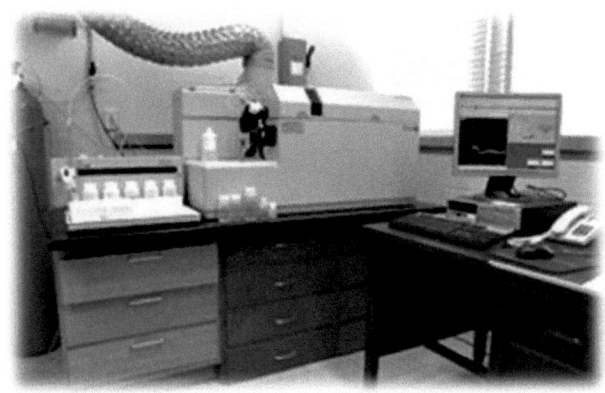

Figure 4.1. Mass spectrometer with inductively coupled plasma ISP-MS Agilent 7500

For the analysis of trace elements and heavy metals in braga whit the aid of ISP-MS technique 5 ml of the sample were extracted in a crucible, using a syringe, the mixture having a neutral-slightly acidic pH (between 5.5 – 6) and introduced in an oven at a temperature of 200°C, maintaining up to completely evaporation on the liquid. The method is based on the principle of calcination of the sample at 450°C with gradual increase of the temperature and the dissolving of ash with 2 ml HNO$_3$ 65%. The mixture is brought to the mark with distilled water in a 25 ml graduated flask, after which the sample is subjected to analysis at the ISP-MS, the electrical component generating corresponding results which are processed by specific calculation techniques.

The calibration is carried out with a multi-element solution (Li, Be, B, Na, MG, Al, K, as well as, V, Cr, Mn, Fe, Fe, Co, Ni, Cu, Zn, GA, As, se, Rb, Sr, Ag, Cd, Cs, BA, are you, Pb, U) purchased from a specialized company and the concentration of each component was correlated with the 10 mg/l and then the quantity of multi-element in the distilled water sample used for dilution was determined.

4.2. Result and Discussion

According to the Article 63 of Order No. 611 of 3rd April 1995 regarding the approval of hygiene norms of food and sanitary protection thereof, published in M.O.

No. 59bis of 22nd of March 1996 braga must meet the following quality requirements: dry matter in refractometric degrees at 20°C min. 7, acidity in lactic acid not exceeding 0.65g/100ml of the product. In case when altered raw materials, synthetic sweetening substances, emulsifiers or preservatives are used, it is not permitted for human consumption (www.legex.ro)

Having as objective to identify of trace elements and heavy metals in braga, after having interpreted the results obtained from ICP-MS analysis and after having made the calculations, a graph in figure 4.2. is generated, representing the variation in concentration of trace elements.

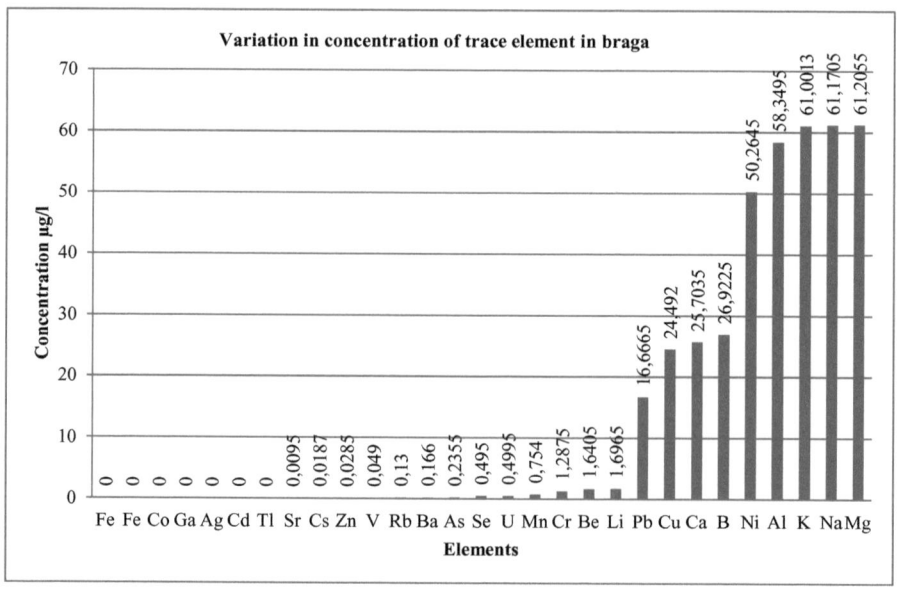

Figure 4.2. The plot of the concentrations of trace metals and minerals in braga

Thus, it has been demonstrated that braga contains minerals such as: Mg, Na, K, As, Mn, Zn, magnesium registering the highest concentration, respectively 61,2055 µg/l and zinc the lowest one, respectively 0,0285 mg/l.

As regards the content of heavy metals, the data obtained from the analysis demonstrated the presence of Al, Ni, Pb, Cu, Cr, U, I, aluminum and nickel recorded increased values of concentrations. The analysis of the results obtained which comply with those on current rules regarding food safety (Law 311/2004 and STAS 6362/85) shall reflect the fact that braga is a proper product, the heavy metals identified having a concentration value that does not exceed the maximum permissible concentration limits.

Having in view the fact that in the preparation of beverage dishes and utensils containing metals were not used, the most likely origin of these heavy metals traces would be explained by the raw materials used or the soil where cereals were grown.

CONCLUSION

The returning to the „product of the past" and capitalizing of our ancestors' recipes might be a healthy choice among all consumers nowadays and the cereal fermented beverages, such as braga constitutes the most effective and accurate option to everything that is being promotes by the contemporary market in term of soft drinks.

The nutritional value and human health benefits manifested by fermented beverages remain undisputed. While some food producers believes that the promotion of such drinks is less attractive and would not turn into a promising market, many researchers argue that the marketing of these healthy drinks is provided by the local traditions of areas where the nutrient valences are well known.

Moreover, the nutritional treasure contained by these products, along with appropriate consumer information brings extra confidence and it is the most reliable proof that such products may be daily placed on the table by each of us.

While many consumers are very aware of the nutritional qualities of food, especially the organoleptic characteristics of products, most of them are often easily influenced by the colors of container and lack of labeling information, opting for a „pleasure" which could harm them without awareness. Despite its not very attractive aspect, braga is a healthier option to what the market offers today. Its nutritional qualities are provided by an amount of nutrients such as vitamins, minerals and probiotics, conferring it the status of a high quality functional food meant for any group age. The fact that braga has a mineral content and heavy metals that ensure its innocuousness represents an argument that this drink is safe for consumers, and moreover, it bring extra energy thanks to its special qualities.

I believe that this cereal beverage must be properly valorized, not only in terms of its nutritional content, but also in terms of technology, encouraging its production at industrial level. Thus, the valuable recipes of our ancestors, enriched with the results of various studies on this drink could be the base for a broader approach to this miraculous liquor, beeing available on every market or store shelf.

The perspective regarding the traditional fermented drinks shows that there must be a consensus on what the microbiota of the natural product consists of, to describe the contribution of each microorganism on the final composition of the beverage and of the relationship between microorganisms (eg between bacteria and yeast populations). All these issues can be elucidate by technological means that advanced increasingly, in this way making possible a theoretical base easily to address to. The role of these traditional drinks is to diversify and develop new products to the market, taking into consideration especially their effects on consumers' health. However, in this way the collaboration between industry and academia might be developed, aiming at with more attention the satisfy of customers' needs in a more effective mode (Alan, J. Marsh et al., 2014).

REFERENCES

1. Akpinar-Bayizit, A., Yilmaz-Ersan, L., Ozcan, T. (2010), *Determination of boza's organic acid composition as it is affected by raw material and fermentation*, International Journal of Food Properties, 13, 648–656;

2. Alan, J. Marsh, Colin, Hill, Paul R., Ross, Paul D., Cotter, (2014), *Fermented beverages with health-promoting potential: past and future perspective,* Trends in Food Science and Technology, 38, pp. 113-124;

3. Angelov, A., Gotcheva, V., Kuncheva, R., Hristozova, T., (2006), *Development of a new oat – based probiotic drink*, International of Food Microbiology, 112, 75-80;

4. Arici, M., Daglioglu, O., (2002), *Boza: a lactic acid fermented cereal beverage as a traditional Turkish food*, Food Reviews International, 18, 39–48;

5. Bârcă, Adriana, *Gastronomy and gastrotehnie - course notes,* Suceava, 2011;

6. Caplice, E., Fitzgerald, G.F., (1999), *Food fermentations: role of microorganisms in food production and preservation,* International Journal of Food Microbiology, 50, 131–149;

7. Cara, Daniela, Vlad, Gheorghiță, (2014), *Food safety - European context,* Certind Magazine, No.1;

8. Cosansu, S., (2009), *Determination of biogenic amines in a fermented beverage, boza,* Journal of Food Agriculture and Environment, 7, 54-58;

9. Crăciun, Veronica Isabela, (2011), *Studies on biologically active substances used as food additives and nutrients to improve quality and food security - Summary thesis,* University "Lucian Blaga" of Sibiu;

10. Culea, M., (2008), *Mass spectrometry. Principles and Application,* Publishing Risoprint, Cluj-Napoca;

11. Dabija, Adriana, (2010), *Biotechnology in the food industry fermentative,* Publishing Risoprint, Iași;

12. Gassem, A.A. Mustafa, (2002), *A microbiological study of Sobia: a fermented beverage in the Western province of Saudi Arabia*, Department of Food Science and Nutrition, College of Agriculture, King Saud University;

13. Gotcheva, V., Pandiella, S. S., Angelov, A., Roshkova, Z. G., Webb, C. (2000). *Microflora identification of the Bulgarian cereal – based fermented beverage boza. Process Biochemistry, 36*, 127–130.

14. Gotcheva, V., Pandiella, S. S., Angelov, A., Roshkova, Z., Webb, C., (2001), *Monitoring fermentation of the traditional Bulgarian beverage boza*, International Journal Food Scence an Tehnology, 36, 129-134;

15. Gutt, Sonia, Gutt, G., (2005), *Instrumental Analysis - spectroscopy*, Publishing Universităţii Suceava;

16. Hangioglu, O., Karapinar, M., (1997), *Microflora of Boza, a traditional fermented Turkish beverage*, International Journal of Food Microbiology, 35, 271–274;

17. Ion, V., (2010), *Fitotehnie*, Iaşi;

18. Iordache, A. M., (2011), *The study of complex systems with biomedical and environmental applications - abstract thesis*, University Babeş-Bolyai, Cluj-Napoca;

19. Jantschi, L., (2004), *Chimie Fizică Analize Chimice şi Instrumentale*, Academic Direct, Cluj-Napoca;

20. Jaskari, J., Kontula P, Siitonen, A., Jousimies-Somer, H., Mattila-Sandholm, T., Poutanen, K., (1998), *Oat-beta-glucan and xylan hydrolysates as selective substrates for Bifidobacterium and lactobacillus strains*, Applied Microbiology and Biotechnology, 49, 175-181;

21. Koehler, P., Wieser, H., (2013), *Chemistry of Cereal Grains – Chapter 2*, German Research Center for Food Chemistry;

22. Köse, E., Yücel, U. (2003), *Chemical composition of Boza*, Journal of Food Technology, 1, 191–193;

23.LeBlanca, J.G., Todorov, S.D., (2011), *Bacteriocin producing lactic bacteria isolated from Boza, a traditional fermented beverage from Balkan Peninsula – from isolation to application*;

24.Li, J., Zhang, W., Wang, C., Yu, Q., Dai, R., Pei, X., (2012), *Lactococcus lactis expressing food-grade β-galactosidase alleviates lactose intolerance symptoms in post-weaning Balb/c mice*, Applied Microbiology and Biotechnology;

25.Maarse, H., Visscher, C.A., Willemsens, L. C. and Boelens, M. H., (1992), *Volatile compound of Food: Qualitative and Quantitative Data.* Zeist, Netherlands, TNO Biotehnology and Chemistry Institute;

26.Oliveira, José M., Genisheva, Zlatina, Vilanova, Mar, (2009), *Fermented Beverages – process technology and management, volatile compounds, instrumental methods of analysis*;

27.Rafter, J. (2004), *The effects of probiotics on colon cancer development,* Nutrition Research Reviews, 17, 277–284;

28.Rusu, Teodora Emilia., (2011), *Comparative study of biomarkers for quality, safety and authenticity Romanian traditional distilled spirits - abstract thesis,* University of Agricultural Sciences and Veterinary Medicine, Cluj-Napoca;

29.Salovaara, H., Simonson, L., (2003), *Fermented cereal-based functinal foods,* Handbook of Fermented Food and Beverage;

30.Siminiuc, Rodica, Coşciug, Lidia, Bulgaru, Viorica, (2004), *Changing the amino acid content after skinning and hydrothermal treatment of the soriz grains;*

31.Soceanu, Alina Daria, (2009), *Physicochemical and analytical study of pollutants from plants - thesis,* Bucureşti;

32.Tănăselia, C., (2013), *Applications of mass spectrometry with inductively coupled plasma in the analysis of heavy metals in environmental samples - abstract thesis,* University Babeş-Bolyai Cluj-Napoca, Faculty of Physics;

33.Todorov, S.D., (2010), *Diversity of bacteriocinogenic lactic acid bacteria isolated from boza, a cereal-based fermented beverage from Bulgaria,* Food Control, 21, 1011–1021;

34. Todorov, S.D., Botes, M., Guigas, C., Schillinger, U., Wii, I, Wachsman, M.B., Holzapfel, W.H., Dicks, L.M.T., (2008), *Boza, a natural source of probiotic lactic acid bacteria*;

35. Todorov, S.D., Dicks, L.M.T., (2005), *Pediocin ST18, an anti-listerial bacteriocin produced by Pediococcus pentosaceus ST18 isolated from boza, a traditional cereal beverage from Bulgaria*, Process Biochemistry, 40, 365–370;

36. Todorov, S.D., Dicks, L.M.T., (2006), *Screening for bacteriocin-producing lactic acid bacteria from boza, a traditional cereal beverage from Bulgaria. Comparison of the bacteriocins,* Process Biochemistry, 41, 11–19;

37. Todorov, S.D., Holzapfel W.H., (2014), *Traditional cereal fermented foods as sources of functional microorganisms*, Universidade de São Paulo, Brazil;

38. Vasudha, S., Mishra, H.N., (2013), *Non dairy probiotic beverages*, International Food Research Journal 20(1), 7-15;

39. Von Mollendorff, J. W., Todorov, S. D., Dicks, L.M.T., (2006), *Comparison of bacteriocins produced by lactic acid bacteria isolated from boza, a cereal-based fermented beverage from the Balkan Peninsula,* Current Microbiology, 53, 209–216;

40. Wang, Chung-Yi, Sz-Jie Wu, Y.T.S., (2013), *Antioxidant properties of certain cereals as affected by food-grade bacteria fermentation*, Journal of Bioscience and Bioengineering;

41. Yegin, S., Uren, A. (2008), *Biogenic amine content of boza: a traditional cereal-based, fermented Turkish beverage*, Food Chemistry, 111, 983–987;

42. Zorba, M., Hancioglu, O., Genc, M., Karapinar, M., Ova, G. (2003), *The use of starter culture in the fermentation of boza, a traditional Turkish beverage,* Process Biochemistry, 38, 1405–1411.

43. *** technical specification *SC Comalina SRL*

On-line references

1. *** http://www.121.ro/fitness-si-dieta/braga-sau-boza-o-bautura-racoritoare-uitata-35368

2. *** http://adevarul.ro/locale/galati/braga-galaticea-mai-romania-amenintata-falsi-producatori-comerciantinecinstiti-1_5176eedd053c7dd83f3d8226/index.html#photo-head

3. *** http://www.bauturinaturale.ro/braga-382.html

4. *** http://www.culinar.ro/articole/culinar-prin-lume/afla-totul-despre-braga/202/2170/

5. *** http://dambovita.net/index.php/sanatate/item/414-braga-istoria-unei-bauturi-aproape-disparuta-din-romania

6. *** www.gds.ro

7. ***http://www.hurriyetdailynews.com/the-ottomans-favorite-winter-drink---boza.aspx?PageID=238&NID=77350&NewsCatID=438

8. *** http://www.legex.ro/Ordin-611-1995-7825.aspx

9. *** http://metropotam.ro/La-zi/A-fost-odata-ieftin-si-autohton-Braga-art2283169057/

10. *** http://nopr.niscair.res.in/bitstream/123456789/8026/1/NPR%205%286%29%20422-427.pdf

11. *** http://www.producatorbraga.ro/despre-braga/

12. ***http://www.stiri-evenimente.ro/lifestyle/ieftin-ca-braga-si-sanatos-sa-ne-amintim-de-bautura-noastra-balcanica-braga.html

13. *** http://www.tandfonline.com/doi/pdf/10.1081/FRI-120003416#.VP71btKsXOt

14. *** www.wikipedia.en

15. *** http://ro.wikipedia.org/wiki/Brag%C4%83

16. *** www.springer.com

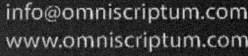